Israel C. Russell

Lakes of North America

A reading lesson for students of geography and geology

Israel C. Russell

Lakes of North America
A reading lesson for students of geography and geology

ISBN/EAN: 9783337322182

Printed in Europe, USA, Canada, Australia, Japan

Cover: Foto ©berggeist007 / pixelio.de

More available books at **www.hansebooks.com**

LAKES

OF

NORTH AMERICA

A READING LESSON

FOR STUDENTS OF GEOGRAPHY AND GEOLOGY

BY

ISRAEL C. RUSSELL
PROFESSOR OF GEOLOGY, UNIVERSITY OF MICHIGAN

Boston, U.S.A., and London
PUBLISHED BY GINN & COMPANY
1895

ANN ARBOR, MICHIGAN,
April 12, 1894.

GROVE KARL GILBERT,

U. S. GEOLOGICAL SURVEY,
WASHINGTON, D. C.

MY DEAR SIR: —

It is now fourteen years since you first guided my footsteps to the beaches of Lake Bonneville and pointed out the striking contrasts in the sculpturing of the mountains above and below the horizon to which that ancient sea flooded the now desert valleys of Utah. For several years after the survey of Utah's former lake was completed, you directed my studies of the basins of similar lakes in Nevada, California, Oregon, and Washington; and through your advice and suggestions I was enabled to see many things that otherwise might have escaped notice.

While writing this little book, which so inadequately describes some of the most interesting events in the later geological history of North America, I have made more use than I could well acknowledge of your volume on Lake Bonneville and of your more general discussion of the Topography of Lake Shores — books that are numbered among the classics of American geology.

As a partial acknowledgment of this accumulated indebtedness, I beg to be allowed to dedicate this book to you.

I remain, very respectfully,

ISRAEL C. RUSSELL.

PREFATORY NOTE.

A LARGE portion of the facts pertaining to the lakes of North America, presented in this book, were gleaned by the writer during thirteen years' geological work for the National Government, and are recorded principally in the publications of the U. S. Geological Survey. The facilities for exploration afforded by my connection with Government surveys enabled me to visit various parts of the United States, inclusive of Alaska, and to observe many phases in the topographical development of our continent.

The publications of the U. S. Geological Survey, and of several State surveys, also contain the records of observations by others, relating to the subject here treated, which have been freely used. It is hoped that this popular presentation of a small part of the results of the various surveys referred to will serve to direct attention to the rich and varied store of information contained in the reports of my colleagues and fellow-workers.

Besides the publications of official surveys, many papers relating to the subject here discussed have appeared in journals, proceedings of scientific societies, etc., to which references may be found in footnotes in this volume.

The origin of lake basins and the history of the great cycles in the development of the relief of the land to which they pertain, have been discussed especially by Professor W. M. Davis, of Harvard University. Professor Davis has also read the manuscript of this book and kindly given me the benefit of his criticisms and suggestions.

I. C. R.

INTRODUCTION.

LAKES have their birth and death in the topographic development of the land. A certain class form a characteristic feature of lands recently elevated above the sea; others belong with the earlier stages or youth of streams; while still others appear during maturity or in the old age of the rivers to which they owe their origin. Lakes of a different type are associated with modifications of topography due to glacial and to volcanic agencies, and to movements of elevation and depression in the earth's crust.

Lakes, like mountains and rivers, have life histories which exhibit varying stages from youth through maturity to old age. The span of their existence varies as do the lives of animals and plants. In arid regions they are frequently born of a single shower and disappear as quickly when the skies are again bright; their brief existence may be said to resemble the lives of the Ephemera. Again, the conditions are such that lakes perhaps hundreds of square miles in area, are formed each winter and evaporate to dryness during the succeeding summer; these may be compared with the annual plants, so regular are their periods. Still others exist for a term of years and only disappear during seasons of exceptional aridity; but the greater number of inland water bodies resemble the Sequoia, and endure for centuries with but little apparent change. So long are the lives of many individuals that human history has recorded only slight changes in their outlines, but to the geologist even these are seen to be of recent origin and the day of their extinction not remote.

The tracing of the life histories of lakes and the recognition of the numerous agencies that vary their lives and lead to their death, gives to this branch of physiography one of its principal charms.

Lakes are also expressive of climatic conditions. In humid regions they usually overflow, are fresh, and vary but slightly in area or in depth, from season to season, and from century to century. In arid lands they are frequently without outlets and consequently alkaline and saline, and fluctuate in sympathy with even the minor changes in their climatic environment.

The history of a lake begins with the origin of its basin and considers among other subjects the movements of its waters, the changes it produces in the topography of its shores, its relations to climate, its geological functions, its connection with plant and animal life, etc. It is in this general order that the lakes of North America are considered in the present volume. The standpoint from which the subject is treated is that of the geologist and geographer, its relation to man being left to the archaeologist and the historian.

CONTENTS.

—◦—

INTRODUCTION.

CHAPTER I.

ORIGIN OF LAKE BASINS.

CHAPTER II.

MOVEMENTS OF LAKE WATERS AND THE GEOLOGICAL FUNCTIONS OF LAKES.

CHAPTER III.

TOPOGRAPHY OF LAKE SHORES.

ILLUSTRATIONS.

LAKES OF NORTH AMERICA.

CHAPTER I.

ORIGIN OF LAKE BASINS.[1]

DIFFICULTIES arise in classifying lake basins, similar in character to those met with when a systematic discussion of glaciers, rivers, mountains and other features of the earth's surface is attempted. That is, there are no natural groups separated by hard and fast lines, into which they naturally fall. Certain types may be selected, however, answering to genera among plants and animals, about which most lakes may be grouped. In selecting these types we are guided by their mode of origin, and are thus led to an incomplete genetic classification, based on the natural agencies which produce depressions in the earth's surface.

Depressions on new land area. — On lands recently elevated above the sea or left exposed by the evaporation or drainage of inland water bodies, there are usually inequalities, and water frequently collects in the depressions and forms lakes. There are comparatively few lakes of this type in North America, for the reason that large portions of our coasts are sinking and new land areas are rare. The lakes of Florida, however, are good examples of this class. They are surrounded by marine rocks of recent origin, and are but slightly elevated above the sea. In fact, all of the topographic features of Florida indicate immaturity. The luxuriant vegetation of the southeastern coastal plain, masks the slight inequalities of the surface, and, by clogging the slack drainage, leads to a greater

---

[1] This subject has been discussed by numerous writers, and has led to controversies not yet ended. The most extended and most systematic treatment that it has received may be found in an essay by W. M. Davis "On the classification of lake basins," in Boston Soc. Nat. Hist., Proc., vol. 21, 1882, pp. 315–381. The numerous references given in this paper constitute the best bibliography of the subject available. An important supplementary paper by the same author is republished as an appendix of the present volume.

expansion of the lakes than would appear if the land was barren. The wealth of vegetation tends also to preserve the original barriers from erosion. About the southern shore of Hudson bay there is another area recently abandoned by the sea, on which there are lakes, but this region is so little known that it cannot be pointed to with confidence as a case in point. In the Great Basin, as the vast area of interior drainage between the Sierra Nevada and Rocky mountains is termed, there are many lakes, some of them of large size, which occupy depressions in the surfaces of sedimentary deposits left exposed by the evaporation of much larger Pleistocene water bodies. Great Salt lake and Sevier lake, Utah, occupy the lowest depressions in valleys formerly flooded by the waters of a great inland sea to which the name Lake Bonneville has been applied. Pyramid, Walker and other lakes in Nevada, occur in valleys which are deeply filled with the sediment of another ancient water body named Lake Lahontan. In these instances, however, and in many others of similar character throughout the Arid Region, the positions of the present lakes on the approximately level floors of desert valleys have been partially determined by recent movements of large blocks of the earth's crust adjacent to lines of fracture, and by the unequal deposition of alluvial material swept out from mountain valleys and deposited on the adjacent plain. These recent changes have modified the character of the basins now occupied by lakes, but essentially they are depressions on new land areas, and form the most typical examples of their class that can be found in this country.

There are new land areas about the borders of the Laurentian lakes, which have been left exposed by the recession of still greater lakes that occupied the same basin at a comparatively recent date, and also in the region drained by Red river in Minnesota and Canada, formerly flooded a vast lake named in honor of Louis Agassiz. Along some of our rivers, also, which flow through ancient valleys now deeply filled, there are narrow areas of new land, similar to the recently exposed borders of the Laurentian lakes. In all of these instances, however, the lakes formed in the inequalities of the surface are small and of little importance.

Lakes on new land areas are surrounded by topographic forms expressive of youth, and are themselves evidence of topographic immaturity. When drainage is established on such areas the basins are soon emptied. The lives of lakes of this class, as is the case with all terrestrial water bodies, depend largely on climatic conditions. They may continue longer in one region than in another, but in the ordinary course of topographical development are transient features. In humid regions they are drained more quickly than where the rainfall is small. They are fresh or saline according as they overflow or are without outlet.

On old land areas where the streams have reached maturity or old age, the inequalities of the surface due to the accidents of original deposition are removed, and lakes of the class here considered are absent. This is shown in a striking manner by contrasting Florida with the adjacent Appalachian region. In the former, lakes are abundant, and their surroundings give abundant evidence of recent origin ; in the latter, the topographic forms as well as the terranes from which they have been carved, bear the stamp of antiquity.

Lands that have been subjected to intense glaciation, or have. received a covering of glacial deposit, are essentially new land area, and bear evidence of topographic youth ; but the lakes characteristic of such rejuvenated lands will be considered in advance in connection with other results of glacial action.

Basins due to atmospheric agencies. — The weathering of rock surfaces progresses unevenly, on account of varying hardness and the varying degree to which they yield to chemical changes. This is noticeable particularly on granitic areas, as granite is especially prone to disintegration, and produces uneven surfaces when weathered. The tendency to decay unequally, as weathering progresses, probably exists in all rocks ; and it is to be expected that hills and hollows would result for the action of the atmosphere on any variety of deposit, especially if marked variations occur in its texture and composition. This tendency is most easily detected when the bedding is nearly horizontal, and large sheets of nearly level strata are exposed to the sky.

The products of weathering are removed by water in solution and in suspension, and are blown away by the wind. When removed by water, the formation of basins is checked by the cutting of outlets. When carried away by the wind, depressions known as " wind-erosion basins " are left.[1] These are basins of excavation or true rock basins, and in this respect resemble depressions eroded by glaciers. Some observers have concluded that many of the rock basins commonly ascribed to glacial

[1] Numerous examples of shallow, saucer-shaped depressions in shale, due to the action of the wind on areas bare of vegetation, in the southeastern part of Colorado, have recently been described by G. K. Gilbert. Jour. of Geol., vol. 3, 1895, pp. 47–49.

action, are wind erosion basins or areas of pronounced rock decay, from which glaciers have removed the loosened material without deeply abraiding the unweathered rock beneath. The mode of origin of rock-basins is still a matter of controversy, but it seems evident to the writer, not only from reading the various views advanced by others, but also from personal observation in many lake regions, that rock basins have been formed by each of the agencies mentioned as well as by a combination of the two. The formation of basins by ice erosion and by chemical solution might be included among the results of atmospheric action, but under the classification here adopted they fall in different categories.

Atmospheric agencies also lead to the formation of basins by deposition ; as for example, when sand is drifted into dunes. Drifting sand frequently travels across the country for scores of miles in the direction of the prevailing winds, and sometimes obstructs valleys so as to cause lakes to form. The best illustration of this occurrence known to the writer, is in the central part of the State of Washington. The drainage of one of the deep narrow valleys known locally as "Coulees," which trench the Great Plain of the Columbia, has been obstructed by immense sand dunes, so as to form a dam and retain the water of Moses lake.[1] Below the dam of drifted sand there are several springs fed by lake waters percolating through the obstruction. These serve to keep the waters of the lake fresh. The springs below the sand drifts unite to form Alkali creek, which in winter sometimes has sufficient volume to reach the Columbia, but in summer suffers from evaporation, and terminates in a series of alkaline pools.

Drifting sand may lead to the destruction of a lake as is illustrated by an example in western Nevada. The branch of Truckee river, supplying Winnemucca lake, is partially obstructed by wind-blown sand, and a struggle for supremacy between the river and the encroaching dunes is in progress. Should the sands prevail and a dam be formed, the water supply of Winnemucca lake would be diverted to Pyramid lake, and its basin would soon become desiccated.

Volcanic dust is carried great distances by air currents, and might accumulate in a valley so as to obstruct its drainage. No lakes, retained by dams of this nature, are known on this continent, although thousands of square miles in the western part of the United States were covered, in Pleistocene and recent times, to a depth of many feet with fine volcanic

[1] I. C. Russell, "Geological Reconnoissance in Central Washington," U. S. Geol. Surv. Bulletin, No. 108.

deposits, which in some instances have assisted other agencies in producing inequalities of the surface.

Basins due to aqueous agencies. — In this class of basins there are two important subdivisions : *a*, basins due to the action of streams, and *b*, basins due to the action of waves and currents. In each subdivision, but more especially in the first, there are basins formed by excavation and basins due to deposition, or basins due to destructive and to constructive agencies. Frequently the two processes have united in the formation of a single depression.

a. Basins formed by streams. — The drainage of new land areas, especially in humid regions, soon obliterates the depression due to the original inequalities of the surface, as already explained; but other basins resulting from the action of the streams themselves are formed.

When the topography of a young land area is yet immature, and more especially when the elevation is considerable and the climate humid, the even flow of the draining streams is apt to be interrupted by rapids and water-falls, at the bases of which excavation is accelerated and depressions formed. The deepening of such portions of stream-beds, results principally from the friction on their bottoms and sides, produced by sand and stones moved by the swift currents. Some distance below falls and rapids, the current usually slackens, and the waters deposit a portion of their load. A basin of this character is now being excavated below Niagara falls, and other examples may be seen in the channels of many mountain streams. Even on old land areas like the southern portion of the Appalachian region, where the streams are engaged in cutting down synclinal table-lands in which hard and soft strata alternate, small basins of the character here referred to are of common occurrence. Should a stream channel in which such inequalities have been produced be abandoned as a line of drainage, the basins would be transformed into lakes.

The best example of a lake basin of considerable size formed at the base of a water-fall, that has come under the writer's notice, is in the Grand Coulee, near Coulee City, in the State of Washington. The Columbia river now skirts the northern and western borders of the vast lava-covered region known as the Great Plain of the Columbia, or more familiarly as the "Big Bend country," but in Pleistocene times its present course was obstructed by glaciers which descended from the mountains to the north, and it was forced to cut across the Big Bend through a series of deep cañons in the lava. Its temporary course was through

Grand Coulee, and near the present site of Coulee City, it plunged over
a precipice about two hundred feet high, and formed a cataract of the
nature of Shoshone falls, Idaho, but rivaling Niagara in grandeur.   Two
basins were excavated in the rocks at the base of the falls, which were
left as lakes when the glaciers retreated and the Columbia returned to its
old channel.   These lakes still exist although desert shrubs grow on the
brink of the precipice over which the waters of the flooded and ice-laden
river previously thundered.   Each of the lakes is by estimate a mile
long and half a mile broad, and of considerable depth, as is shown by the
dark blue color of their waters when seen from the crest of the encircling
cliffs.[1]

The deeper positions of stream-channels excavated during floods, may
be transformed into lakes when the waters subside or when the course of
a stream is changed.   This is shown by the temporary ponds remaining
in many humid countries during droughts when water no longer flows
through the customary surface channels, but is more common in arid
regions where the streams are subjected to still greater fluctuations.

The basins just described are formed principally by excavation, those
noted below are due to deposition.

In regions of rapid erosion, a high grade and consequently rapid
tributary, may bring to a sluggish trunk stream more detritus than it is
able to carry away.   When this happens, the main stream is more or less
completely obstructed, and lakes may result.   Basins of this nature occur
in the steep-walled valleys of the Sierra Nevada and Rocky mountains,
and are to be expected wherever streams have cut back their trenches far
into an upland and receive high-grade tributaries.

The alluvial cones about the bases of mountains in the Arid Region
are frequently several miles in radius, and have a thickness near the
mouths of the gorges from which the material forming them was discharged, of two or three thousand feet or more.   When such deposits are
formed on the opposite side of a valley only a few miles across, they may
unite one with another so as to form transverse ridges and give origin to
basins.   Alluvial cones are especially conspicuous in regions where the
drainage in the valleys is weak or entirely wanting, thus favoring the
formation of basins in the manner just described.   Lake Tulare, in
southern California, may be cited as an example, as it is retained on a
broad alluvial plain by material swept out by torrents from cañons in the

<hr>

[1] I. C. Russell, "Geological Reconnoissance in Central Washington," U. S. Geol. Surv.
Bulletin, No. 108.

Sierra Nevada. In regions where the conditions are most favorable for the growth and preservation of alluvial cones, there is but little rain-fall, and the material deposited in the valleys is apt to be porous and of such a character that it absorbs water readily; for this reason lakes may be absent and the land remain desert-like and arid although basins exist.

A lack of close adjustment in the transporting power of streams may sometimes be observed even in humid countries, and in regions of mild relief. As described by G. K. Warren,[1] the excess of material brought by Chippeway river to the Mississippi, obstructs the main stream so as to cause an expansion of its waters known as Lake Pepin. An approximation to the same conditions occurs where Wisconsin river and Illinois river join the "Father of Waters"; but in these instances it is only in the low water stages that the ponding becomes conspicuous. A tendency in the same direction was noted by J. W. Powell while making his adventurous journey through the cañon of the Colorado; dangerous rapids were encountered at localities where lateral streams had swept débris into the main channel.

Perhaps the best examples of lakes held by obstructions deposited by lateral streams that can be cited, occur in valleys draining to the Assiniboine, Manitoba. The lakes referred to, are situated in valleys that were cut down to a gentle slope when the abundant drainage of glacial lakes flowed through them; but the weaker modern streams are unable to maintain such a faint grade, and are being silted up where tributaries enter. Long narrow lakes are thus formed above delta-fans built by streams having a higher grade than the main valley.[2]

The separation of lakes Brienz and Thun, Switzerland, has been cited by Davis as an example of the partitioning of a valley by the union of deltas from opposite sides. Interlaken stands on the beautiful alluvial plain thus formed. Several other similar examples in central Europe have been described by various authors.

Lakes retained by the deposits of lateral streams and by alluvial cones, pertain to young and immature streams, and are incident to their work of erosion. As topographic development progresses, these water bodies are obliterated, but when streams reach maturity and old age, lakes of another class appear along their courses.

---

[1] Am. Jour. Sci., vol. 16, 3d ser., 1878, p. 420.
[2] Warren Upham, "Report on Lake Agassiz," Canadian Geol. and Nat. Hist. Surv., Ann. Rep., vol. 4, 1888–89, p. 22 B.

In the case of mature streams that have cut down the seaward portion of their valleys nearly to base-level, that is approximately to the level of the ocean, and where rivers rising in mountainous regions flow across low plains, it frequently happens that the more energetic tributaries towards their head waters bring in more detritus than the gently flowing trunk streams are able to carry, and deposition takes place on their bottoms and over their flood plains. . When the main stream is flooded and inundates its valley, its load is deposited most abundantly on the immediate borders of its channel, and builds up lateral embankments or levees. When this happens, the lateral tributaries joining the main stream in its lower course, may not be able to fill up their valleys as rapidly as the borders of the main river are raised, and are consequently ponded. Many shallow lakes have been formed in this manner along the borders of the large rivers flowing to the Gulf of Mexico. The most conspicuous examples occur along the banks of Red river, Louisiana, where lateral lakes, as has been pointed out by Davis, are arranged along the side of its levees like the leaves on a twig.   ·

In the maturity and old age of rivers, when they meander in broad curves through a wide flood plain, as in the case of the lower Mississippi. the loops are frequently cut off, as shown on Plate 1, and crescent-shaped or "ox-bow" lakes are left. Examples of lakes of this character on a small scale may be seen along the border of many sluggish brooks which traverse deeply filled valleys.

In the formation of low-grade deltas, like those now in process of construction at the mouths of the Mississippi, Nile, Ganges, etc., the waters break through the levees of the main stream during floods, and form branching channels or "distributaries," which in their turn bifurcate in a similar manner, and build up their channels and inundated borders. In such instances low areas are frequently surrounded by embankments, and left as basins containing shallow lakes. Many examples of this occurrence are found on the broad delta of the Mississippi. Of these Lake Pontchartrain is the largest at the present time. Lake Borgne, in the same region, is another example, not yet completed. The delta lands of the Rhine, in Holland, and of other rivers in northern Germany, contain many lakes and swamps of the type here considered. The celebrated Zuyder Zee was formed in part as a delta basin and in part by the·construction of natural embankments adjacent to a low shore. Miniature illustrations of this method of forming basins may be seen on the deltas of many small streams, built in lakes and ponds.

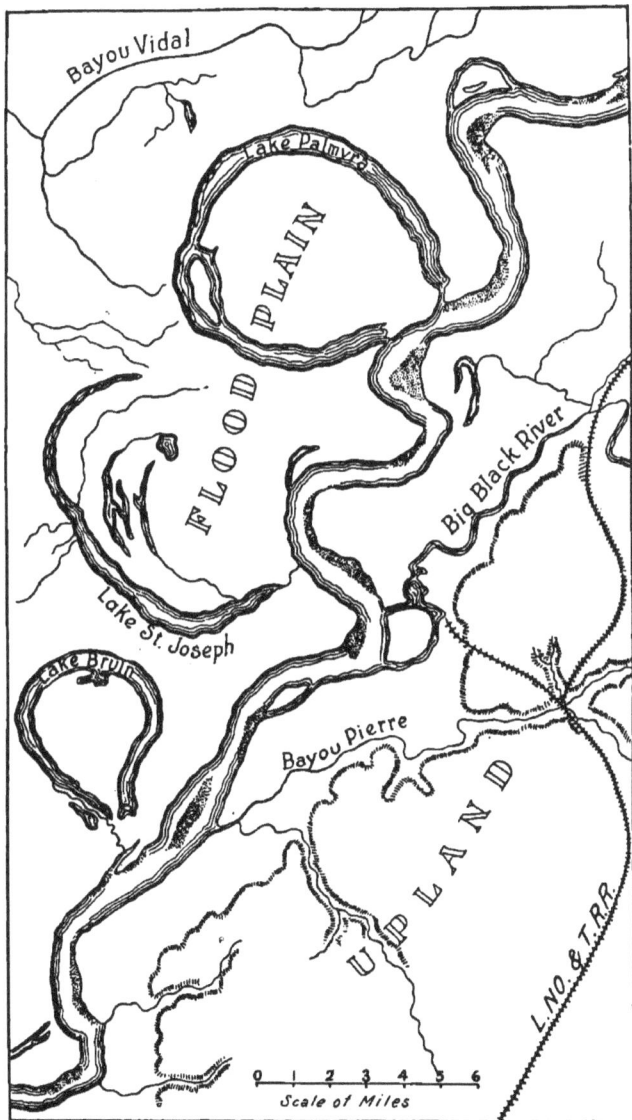

OX-BOW LAKES, LOWER MISSISSIPPI.

ɪ
d
n-
dt
in

spe
llov
n t
10w
ocal
easo

nograj

that they leave but slight if any permanent records. Their waters are so clear that practically no sediments accumulate in them. On continental glaciers, however, such lakes might exist from year to year, and perhaps receive sufficient deposits to leave recognizable records after the ice disappeared. Certain deposits of exceedingly fine, light colored, clay-like material termed *loess*, in the upper Mississippi valley, are believed by some persons who have studied them, to have been accumulated in lakes on the surface of the great ice sheet which formerly covered that region.

When glaciers flow through valleys surrounded by mountains, they sometimes obstruct the drainage of lateral valleys so as to cause lakes to form. The dams in these instances are formed by the ice in the main valleys. The type of this class of lakes is furnished by Märjelen lake, Switzerland. In this instance a lateral valley below the snow line is dammed by Aletsch glacier which flows past its mouth. The lake is variable in area, being sometimes a mile long and at other times completely drained owing to the enlargement of the tunnel beneath the ice dam through which it discharges.

In Alaska there are many lakes of the Märjelen type. About the southern bases of the foot-hills of Mt. St. Elias there are several water-bodies that are held in check by the Malaspina glacier. The largest of these, known as Lake Castani, at the southern end of the Chaix hills, is two or three miles long and a mile broad when at its highest stage, and discharges through a tunnel eight or nine miles long, beneath the ice sheet to the south. The position of this sub-glacial river can be traced by a depression in the surface of the ice, and when above it, the muffled roar of the imprisoned flood can be heard far below one's feet. Of many lakes similar to Lake Castani in the same general region, perhaps the most instructive is one discovered by John Muir, in Stikine valley, British Columbia, near the Alaskan boundary. In this instance a lake about three miles long and approximately a mile broad, and receiving the drainage of five or six residual glaciers, is held in a lateral valley by Toyatte or Dirt glacier, which flows past its entrance. The outlet of the lake is through a tunnel in the ice, which is sometimes enlarged so as suddenly to empty the basin and cause a flood in Stikine river.

The lakes formed when glaciers obstruct the drainage, are variable in size, owing to changes in their draining tunnels, and are frequently emptied, as in instances just cited. The surfaces of these lakes are many times covered with floating ice, which is left stranded when their waters escape. They are unusually turbid with silt brought to them by glacial streams, and leave important deposits to mark their sites when the conditions are no longer favorable to their existence.

The most widely known example of the formation of terraces about the borders of a glacial-dammed lake, is furnished by the Parallel Roads of Glen Roy, on the west coast of Scotland. The origin of these terraces was a fruitful source of controversy for many years ; but the explanation that they are due to the action of the waves and currents of a lake held in a lateral valley by a glacier flowing past its entrance, has finally been accepted as satisfactory.

It is worthy of note, that lakes of the type just described, not only occur in mountain valleys, but also about the ends of mountain spurs projecting into encircling ice sheets, as on the northern border of the Malaspina glacier. The deltas and terraces formed in such lakes may remain in unexpected places, as high up on the side of a mountain, when the retaining glacier is melted.

When the land bordering an ice sheet slopes towards the ice, the escape of the waters formed by the melting of the glacier, as well as streams from the adjacent areas, is checked, and marginal lakes, sometimes of large size, are formed. Two small examples of this class of water-bodies were seen by the writer at the northern base of the Chaix hills, Alaska. During the close of the Glacial epoch, when the ice-sheet occupying northeastern North America was retreating, there came a time when the southern margin of the ice faced a northward-sloping land-surface, and lakes far larger than the present Laurentian lakes, were formed. The largest of these ancient seas, named Lake Agassiz, covered the region in Minnesota and Canada now drained by Red river, and others were formed in the Laurentian basin.

When glaciers melt, the rock surfaces left exposed are frequently planed, grooved and polished. In such instances, the evidences of the friction of the flowing ice and of the sand and pebbles frozen into it, are pronounced and unmistakable. These marks of abrasion are frequently buried and concealed by deposits of débris of various kinds which were transported on the surface of the living glacier, or enclosed in its mass, and left as superficial deposits when the ice melted. In the lower portions of mountain valleys previously occupied by ice streams, and over the outer border of regions formerly covered by continental ice sheets, the deposits of débris are in many instances so abundant that the worn rock surfaces beneath are completely concealed.

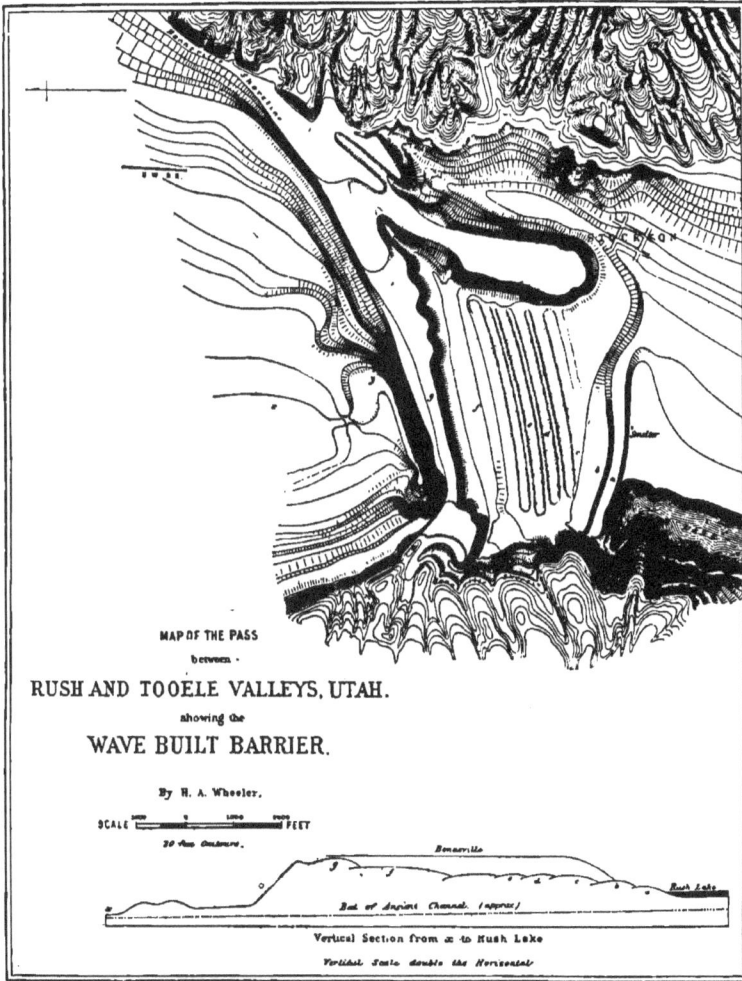

MAP OF THE PASS

between -

RUSH AND TOOELE VALLEYS, UTAH.

showing the

WAVE BUILT BARRIER.

By H. A. Wheeler.

SCALE

STOCKTON BAR UTAH. (AFTER GILBERT.)

Compare with Plate 2.

The study both of living glaciers and of the records left by ancient glaciers has proven that flowing ice both erodes and deposits, and that basins result from each of these processes.

Whether a glacier shall erode its beds or deposit material upon it, seems to depend largely on its grade, and consequently on its rate of flow. In high-grade valleys among mountains formerly occupied by glaciers, the higher and steeper portions of the main avenues of ice drainage, are usually intensely glaciated, and the worn and rounded surfaces are frequently bare of glacial deposits ; but the lower portions of such valleys, especially where they open out on a plain, are almost always heavily covered with morainal material. Not only are moraines deposited in the mouth of the valleys, but sheets of gravel, clay, and boulders are spread over the bottom of the glaciated troughs, showing that the ice-streams in such situations deposited material on the surface over which they flowed.

Above the region of most intense glaciation in lofty mountains there is a zone, embracing the higher summits, where polished and scratched surfaces are rare, and where there is but little débris. This upper region was the site of the névés or snow fields of the glaciers that abraded the rocks at a lower horizon and deposited their loads when the grade decreased and the ice currents were slackened. A similar association of a region of glacial abrasion and an outer zone of glacial deposition, may be recognized in countries formerly covered by continental ice sheets. In the region of most intense glaciation, in the case of both Alpine and continental glaciers, as has been shown by extended observation, there are numerous rock basins, the sides and bottoms of which are polished and striated. A large number of lakes of this character in the Cordilleran region have been examined by the writer, and their study left no doubt that they were due to glacial action. These rock basins are confined to areas of intense glaciation, and are absent from adjacent areas where the conditions are essentially the same, except that glaciers have not passed over them.

It is impossible to point to examples where living glaciers are actually engaged in wearing out rock basins, since their work of abrasion is necessarily concealed; neither is it possible to satisfactorily observe the process by which glaciers polish and striate rock surfaces, yet no student of the subject doubts that these results are produced by moving ice charged with sand and gravel. The nature of the evidence leading to the conclusion that many rock basins are due to glacial abrasion, is of the same character as the evidence from which it is concluded that many smoothed and striated rock surfaces are due to the same agency. The rock basins of the character here referred to, are confined to regions of former glaciation, not only in America but on other continents, and are wanting where other evidences of ice action are absent. The interiors of the basins themselves are smoothed and striated, and bear incontestable evidence that in part at least, they are due to the abrasion of sand-charged ice. These more general considerations are in such harmony with what is known of the work of ice streams, that they carry even more weight than special studies of individual lakes.

Although the evidence leading to the conclusion that many rock basins in glaciated regions are essentially of glacial origin, seems to the writer to be conclusive, it is but just to state that, even after thirty years of ardent controversy, there is still a difference in opinion among geologists and others, in reference to the abrading power of moving ice, and its ability to erode rock basins. The literature bearing on this question is so voluminous that it is impracticable to present even an abstract of it at the present time.[1]

Without considering further the results of the destructive action of glaciers, let us see what is the character of the basin they produce by construction. Fortunately in this connection there is little difference of opinion.

The terminal moraines left by Alpine glaciers in their retreat, frequently form crescent-shaped piles of débris, convex down stream, which act as dams, and retain lakes. Hundreds and probably thousands of examples of lakes held in check by obstructions of this character, exist in the valleys of the Cordilleras, and are common in every formerly glaciated mountainous region. The Twin lakes in the Arkansas valley, Colorado,[2] several small lakes on the west side of Mono valley, California,[3] and numerous sheets of clear water in the Wasatch mountains, Utah, so well known to tourists, are types of this class. Similar lakes occur about the

---

[1] This subject has received special attention since the appearance of a celebrated paper by Ramsay, "On the glacial origin of certain Swiss lakes," Quar. Jour. Geol. Soc., vol. 18, p. 185; but a unanimous conclusion has not been reached, as may be seen by consulting *Nature* for 1893–94. The present status of this interesting controversy is presented in a paper by T. G. Bonney, and accompanying discussions, in the Geographical Journal of the Royal Geographical Society, vol. 1, 1893, pp. 481–504.

[2] F. V. Hayden, U. S. Geol. and Geog. Surv. of the Territories, Ann. Rep., 1874, pp. 47–53. J. J. Stevenson, Explorations and Surveys west of the 100th Meridian ("Wheeler Survey"), vol. 3, 1875, pp. 441–444.

[3] I. C. Russell, U. S. Geol. Surv., 8th Ann. Rep., 1886–87, Pl. 35.

HUMBOLDT LAKE

Elevation 3,943 ft. above the sea.

Soundings in feet.

12

13

10

8

Lahontan Beach.

Lahontan Beach.

Vertical Scale of Profile four times the horizontal.

SCALE OF MILES

1 2 3 4 1 2 3 4

Humboldt R.

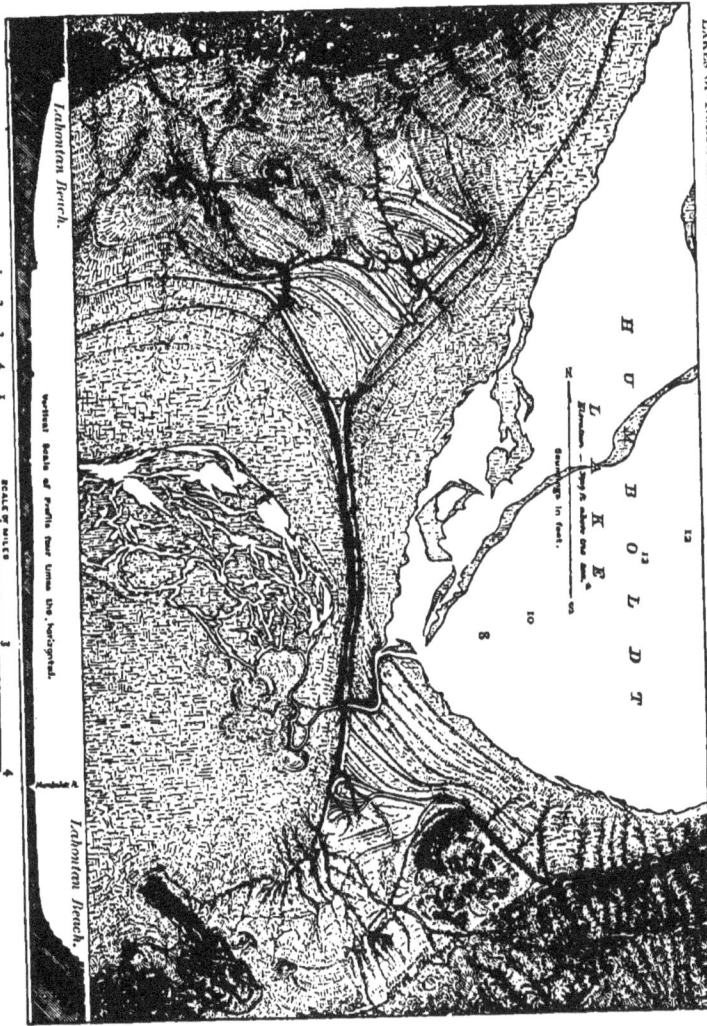

GRAVEL BAR RETAINING HUMBOLDT LAKE, NEVADA.

PLATE 4.

extremities of the existing glaciers of this country, from the High Sierra, California, northward to Alaska. These are retained by moraines, from which the ice has receded within a few years, thus leaving not even the shadow of a doubt as to their mode of origin.

Many of the lakes of Scandinavia and of Switzerland are retained by ancient moraines, as are also in part, the long, deep lakes on the Italian side of the Alps, and draining to the Po. The most striking example of the type of lake here described, however, which has been studied by the writer, is Lake Wakatipu, on the east side of the Southern Alps, New Zealand. This magnificent water body, surrounded on all sides by lofty snow-clad peaks, has many of the characteristic features of lakes Como and Maggiore, and is not second to them in majesty and beauty.

The drainage of mountain valleys, in which moraine-dammed lakes have been formed, is frequently so abundant that stream channels are cut through the obstructions, and the lakes drained. When this occurs, beautiful grass-covered vales or "parks," as they are called in the Rocky mountains, are formed. These charming valleys are quite as beautiful and frequently furnish as great a contrast to the ruggedness of the surrounding scenery, as did the gem-like lakes that preceded them.

In most instances the deep mountain valleys of North America, now occupied by moraine-dammed lakes, were excavated by streams previous to being glaciated, and only served temporarily as avenues for ice drainage. Their main topographic features are due to stream erosion and weathering. Only minor changes such as the smoothing and rounding of their bottom contours, can be ascribed to glacial abrasion.

The general sheets of débris left after the retreat of continental glaciers and by the melting of the expanded extremities of large Alpine glaciers, are usually uneven on account of the manner of their deposition, and abounds in depressions which may hold water. In many instances the lakes originating in this manner are without surface outlets, their surplus water escaping by percolation.

On the formerly ice-covered portion of northeastern North America, the lakes occupying depressions in the general covering of superficial material are so numerous that the position of the southern boundary of the old ice sheet may be approximately traced on a drainage map of the region by noting the southern limit of the lake-strewn portion. The old land surface south of the glacial boundary, is almost entirely free from undrained basins ; and in this, as well as in other respects, presents a striking contrast to the rejuvenated surface of the land to the north.

The lakes occupying depressions on the glacial drift number hundreds of thousands. They vary in size from mere tarns up to splendid water-sheets many square miles in area. In portions of Minnesota, Michigan, and adjacent areas, where the drift is unusually deep, the lakes in irregular depressions on its surface sometimes number a score or more to the square mile. It is estimated that in Minnesota alone, there are not less than ten thousand lakes of this class, besides many swamps and marshes marking the sites of former lakes of the same type, which have become choked with vegetation.

Numerous lakes of the same character as those on the drift of the North-eastern States and Canada, occur about the southern margin of Malaspina glacier, Alaska, in depressions in moraines left by the retreat of the ice within the past few years. These very modern basins, some of which are still occupied in part by the ice of the retreating glaciers, are similar in every way to the basins on the moraine-covered surfaces just referred to, and are surrounded by topography of the same character, thus leaving no room for doubting that each of the two series is due to similar agencies.

When the general sheet of débris left after the retreat of continental glaciers does not completely mask the pre-glacial topography, former valleys are sometimes dammed, and lakes of another type produced. In many instances these lakes are long and narrow, and indicate, to some extent, by their form, the character of the ancient drainage lines they occupy. Again, they may be broad water-bodies, and occupy ancient drainage basins, the outlets of which have been closed. Pre-glacial valleys may be deepened by ice erosions, as well as obstructed, and the two processes may unite to form lakes, as is believed to have been the case in the group of "Finger lakes" in the central part of New York state.[1]

Still another type of lake basins, due to glacial agencies, is found in unconsolidated water-laid material deposited about the borders of ice-sheets. When the stream-borne débris from a glacier is abundant it forms low alluvial cones and sand and gravel plains, which may surround or cover isolated ice masses. When such buried ice masses finally melt a depression is left, and may be water-filled. The borders of such lakes are of loose material which slides into the depression and forms steep banks. The inclination of the enclosing walls depends upon the nature of the material of which they are composed. Broad tracts of sand and

[1] A. P. Brigham, "The Finger lakes of New York," Geographical Soc. Am., Bull., vol. 25, 1893. R. S. Tarr, "Lake Cayuga a rock basin," Geological Soc. Am., Bull., vol. 5, 1894, pp. 339–356.

gravel with hollows of the character just described, scattered over their surfaces, are known as "pitted plains," and find their most acceptable explanation in the hypothesis just suggested.

Lakes Walden and Cochituate, Massachusetts, are believed to be examples of the class of lakes here referred to, and to owe their origin to the melting of ice masses that were either partially or wholly buried in gravel and sand.[1] Lakes of similar character in southern Michigan, where glacial deposits are unusually abundant, might also be cited in this connection. These lakes occupy crater-shaped depressions in the surfaces of gravel and sand plains, of the character that would be expected to result from the burial and subsequent melting of ice masses, in the manner outlined above.

From this brief account of the action of ice in obstructing drainage, it will appear that lake basins are formed not only on account of the damming of streams by the glaciers themselves, but by glacial erosion and glacial deposition; and in still other ways, in connection with the deposits made by streams.

Basins due to volcanic agencies. — Inequalities on the surfaces of lava sheets sometimes give rise to lakes in much the same manner as lakes are formed on the surface of glaciers. Examples of such basins in various stages of extinction, by drainage and sedimentation, occur on portions of the lava plains of Washington and Idaho.

A lava stream may cross a valley so as to obstruct its drainage and cause a lake to form above it, in much the same way as glaciers dam lateral valleys. A large lake was formed in this manner, probably in Pleistocene times, on the Yukon river, Alaska, where it is joined by Pelly river. A series of lava flows there filled the river valley from side to side to a depth of several hundred feet, and formed a dam which retained the waters of the Yukon, and gave origin to a broad water-body known as Lake Yukon.[2] The obstruction has since been cut through along the southern margin of the old channel, leaving a series of basaltic precipices on the right bank of the river.

[1] Warren Upham, Boston Soc. Nat. Hist., Proc., vol. 25, pp. 228–242.

[2] W. M. Dawson, "Report on an exploration in the Yukon district," Canadian Geol. Nat. Hist. Surv., Ann. Rep., 1887–88, p. 132 B.

I. C. Russell, "Notes on the surface geology of Alaska," Geol. Soc. Am., Bull. vol. 1, 1800, pp. 146–148.

C. W. Hayes, "An expedition through the Yukon district," National Geog. Mag., vol. 4, 1802, p. 150.

Another instance of the formation of a lake on account of the filling of a valley by a lava flow, but on a much smaller scale than the example cited above, has been observed by the writer, at the junction of Canadian and Mora rivers, New Mexico. Canadian river, for a distance of perhaps a hundred miles, flows through a steep-walled gorge, in which for a space of several miles, near where Mora river joins it, there is an inner gorge, as indicated in the following cross section :

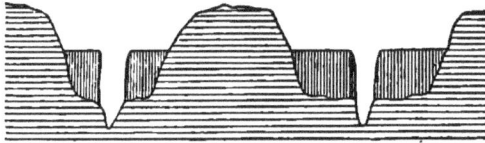

FIG. 1.—CROSS SECTION OF THE CAÑONS OF CANADIAN AND MORA RIVERS, NEW MEXICO
(J. J. STEVENSON).

The valleys excavated by Canadian and Mora rivers were filled to a depth of 400 feet by basalt, as indicated by vertical lines in the section, and were subsequently eroded to a depth of 230 feet deeper than before the obstruction. The lake which existed above the lava flow has been drained, and only indefinite traces of its former presence now remain.[1]

Similar instances of the damming of streams by lava flows, are known on the west slope of the Sierra Nevada, but are also of ancient date. The lakes that were formed have been drained, and their bottoms transformed into grassy valleys.

Two small lakes, held in check by a recent lava stream, now exist at the Cinder cone, near Lassens peak, in northern California. Beneath the lava retaining these lakes there is a sheet of fine lacustral marl and diatomaceous earth, showing that a former lake was partially filled by the molten rock, now hardened into compact basalt.[2]

Another class of lakes due to volcanic agencies, occupy the bowls of extinct craters. These occur in various situations, being sometimes at the summits of high volcanic cones, and again in depressions in broad, featureless plains. The walls enclosing them are sometimes formed of compact lava, but more frequently consist of scoria, lapilli, and so-called ashes, blown out of volcanic vents during periods of violent eruption.

[1] This instructive locality has been described by J. J. Stevenson, in Am. Phil. Soc., Proc., 1880, pp. 84–87.

[2] J. S. Diller. "A Late Volcanic Eruption in Northern California." U. S. Geol. Surv., Bulletin No. 79, 1891.

At Ice Spring buttes, a group of small volcanic craters, near Fillmore, Utah, there is a pool of water in the throat of an extinct volcano, which occupies a depression formed by the recession of the lava that once rose in and partially filled the crater.[1]

The Soda ponds on the Carson desert, near Ragtown, Nevada, occupy lapilli craters, the rims of which rise 20 to 80 feet above the surface of the adjacent plain. The larger pond has an area of 268 acres and a depth of 147 feet, and its surface is 60 feet below the general level of the desert.[2]

A crater similar in character to those holding the Soda ponds, occurs on one of the islands in Mono lake, California, and is occupied by alkaline waters. The water within the crater stands at the same level as the surface of the surrounding lake, a connection between the two being maintained by percolation through the intervening embankment of incoherent lapilli.[3]

One of the numerous craters near San Francisco peak, Arizona, is said to hold a lake at a considerable altitude above the adjacent country. In the summit of Mt. Toulca, Mexico, a deep depression produced by violent eruptions is stated by Davis to have been similarly transformed.

In many volcanic regions in other countries, lakes of this class are known to occur. They are common in Italy, on North Island, New Zealand, and are reported to occur in the Caucasus, on the Solomon Islands, in India, etc. A typical example of a water-filled crater is furnished by Laacher See, on the border of the Eifel, Germany, and has been described and illustrated by Edward Hull.[4]

Still another class of lakes due to volcanic agencies occur where the summits of volcanoes have been blown away by the energy of the confined vapors within ; or when the base of a volcanic pile has been melted so as to cause it to subside into the conduit from which the material composing the mountain was extruded.

It is believed that basins have resulted from each of these processes, but observations on their actual formation are lacking. It is known, however, that volcanic mountains of large size are sometimes literally blown away, as happened in the case of Krakatoa, in 1886.

[1] G. K. Gilbert. "Lake Bonneville." U. S. Geol. Surv., Monograph No. 1, 1800, p. 322.

[2] I. C. Russell. "Lake Lahontan." U. S. Geol. Surv., Monograph No. 11, 1885, pp. 72–80.

[3] I. C. Russell. "Quarternary History of Mono Valley, California." U. S. Geol. Surv., 8th Ann. Rep., 1886–87, p. 373.

[4] "Volcanoes ; Past and Present." Contemporary Science Series, pp. 122–123.

In several volcanic regions there are deep, circular depressions, known as "calderas" or "crater-rings," which are believed to have been formed by the blowing away of the mountains that once existed above them. A somewhat complete series can be established between craters that have been partially broken down by subsequent eruptions, and crater-rings, about which there are in some instances no vestiges of the original craters remaining. There is evidence also in the character of the rocks surrounding crater rings, and in the adjacent topography, which sustains the hypothesis of their violent origin.

Two of the largest calderas yet discovered, occur in Italy, and are occupied by Lago di Bracciano and Lago di Bolsena. As described by J. W. Judd, the first-named is nearly circular, with a diameter of six-and-a-half miles; the second, somewhat less regular, has a length from north to south of ten-and-a-quarter miles, and a breadth of nine miles. The only examples of crater-rings in North America that can be referred to are Gustavila lake, Mexico, of which the writer has been unable to obtain detailed information, and Crater lake, Oregon.

Crater lake has been described by C. E. Dutton,[1] and is considered by him as worthy of a high rank among the wonders of the world. It is situated in the Cascade mountains, in northwestern Oregon, thirty miles north of Klamath lake, at an elevation of 6239 feet above the sea. It is nearly circular, without bays or promontories, as indicated on the accompanying map, Plate 5, and is from five to six miles in diameter. The cliffs of dark basaltic rock encircling it, rise precipitously to heights varying from 900 to 2200 feet, and nowhere offer an easy means of access to the basin within. They plunge at once into deep water, without leaving even a platform at the water's edge wide enough for one to walk on. There are no streams tributary to the lake, and no visible outlet. The waters probably escape by percolation, as the precipitation of the region is in excess of evaporation, and if an escape were not furnished the basin would be filled to overflowing.

Near the southwest margin of the lake, about half-a-mile from shore, a cinder cone, named Wizard island, rises from the water to a height of 645 feet. This cone is regular in form and has a depression in its summit, thus showing at a glance that it is of volcanic origin, and is in fact a miniature crater of eruption. From the base of Wizard island two

[1] Science, vol. 7, 1886, pp. 179–182. Also, 8th Ann. Rep., U. S. Geol. Surv., 1886–87, pp. 157–158.

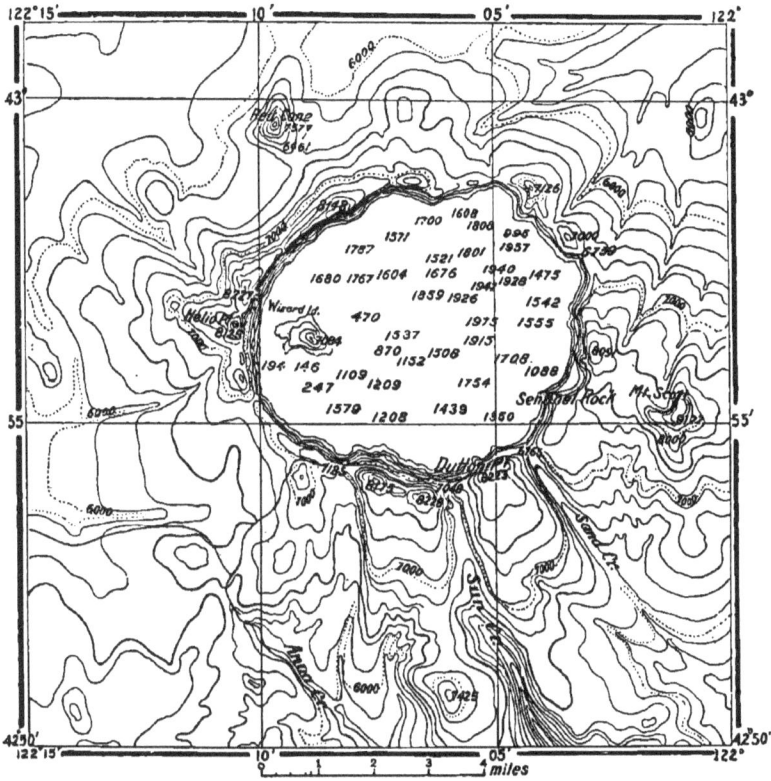

CRATER LAKE, OREGON.   (AFTER U. S. GEOLOGICAL SURVEY.)

Contour-interval 200 feet; soundings in feet; lake surface 6230 feet above sea level.

streams of hardened lava extend outward towards the great walls enclosing the lake, but do not reach them.

The sounding line has shown that Crater lake has a maximum depth of 2000 feet and is the deepest lake now known in North America; its nearest rival being Lake Tahoe. The full depth of the basin measured from the crest of the enclosing cliffs, is from 2900 to 4200 feet.

The rugged slopes encircling the lake as well as the island that seemingly floats on its placid surface, are forest covered, thus softening and rendering picturesque the otherwise oppressive grandeur of the scene.

More remarkable, however, than the unique scenic features of Crater lake, is the story of its origin. The site of the great depression was once occupied by a volcanic mountain which reached far above the highest point on the cliffs now enclosing it, and was probably as conspicuous a member of the sisterhood of mountains of which it formed a part, as any of the neighboring peaks, but the once prominent pile has been removed so as to leave the profound gulf that now fascinates and startles the observer. The character of the sculpturing on the outer slope of the truncated mountain shows that it was eroded, both by streams and by glaciers, before the catastrophe that carried away its summit and left only a hollow stump to mark the site of the ice-crowned peak that formerly gleamed in the sky.

The removal of the summit of the mountain is supposed to have been due to a mighty explosion, similar to that which blew off 5000 feet from Krakatoa; or else that the mountain was melted from within and its summit engulfed so as to leave the depression now partially filled with placid waters. Of these two hypotheses, the second seems to accord best with the observed facts, for the reason that fragmental deposits on the surface of the adjacent country, of the character that would be expected had the summit of the mountain been blown away, have not been recognized. Subsequent to the removal of the summit of the mountain, renewed volcanic energy of a mild character built the crater-island within the crater-ring.

A circular depression in but little disturbed stratified rocks which bears some resemblance to a crater-ring, and which seems likely to furnish the key to the origin of the calderas of Italy and other regions, has recently been discovered in Arizona, about 25 miles southeast of the town of Flagstaff. This unique basin has been carefully studied by G. K. Gilbert, but no account of it from his pen has come under the writer's notice. The observations stated below are mainly from a description of a model of the locality published in the American Geologist.[1]

This "crater" in what is known as Coon butte, is three-fourths of a mile in diameter and its bottom is depressed from 500 to 600 feet below the encircling rim, which rises 150 to 200 feet above the surrounding plains. The surface limestone of the region, elsewhere horizontal, is steeply inclined quaquaversally in the cliffs around the crater; and masses of the same stratum and of an underlying sandstone, are strewn in irregular profusion outward from the crater to the base of the butte, which has a diameter of about two miles. In less amount, the same débris reaches outward on all sides over a nearly circular area to a distance of about four miles. No lava, bombs, lapilli, or other volcanic products, were seen. The formation of this irregular crater-like depression is referred by Gilbert, perhaps provisionally, to a steam explosion.

The occurrence in the vicinity of Coon butte of hundreds of fragments of meteoric iron, up to about a pound in weight, and of several pieces weighing from 20 to 600 pounds, led at first to the thought that possibly a meteorite of great size might have struck this spot, buried itself out of sight and thrown up a crater-like rim. This hypothesis, upon being tested, was abandoned, however, because the volume of the raised rim and of the rock fragments scattered about, was found to correspond very closely with that of the depression below the level of the plain: and for the second reason, that a magnetic survey failed to indicate the existence of any large mass of meteoric iron competent to make such a crater, within at least a depth of many miles. This second objection, however, is now considered of but little weight, since the meteoric fragments found about the crater, although now magnetic, have undergone alterations of a character which seem to indicate that when they first reached the earth they might not have had any or but slight magnetic properties. The changes produced in the surface fragments are due to atmospheric influences, which would not reach a deeply buried body.

The crater-like depression in the summit of Coon butte is without water, for the reason that it is situated in an arid region, but under humid skies would no doubt be transformed into a lake.

The only counterpart of Coon butte as yet discovered, is situated in the central part of the Peninsula of India, some 200 miles northeast of

[1] Vol. 13, 1894, p. 115.

Bombay. This remarkable crateriform lake, known as Lonâs lake, is described by R. D. Oldham [1] as follows :

"The surrounding country for hundreds of miles consists entirely of Deccan trap, and in this rock there is a nearly circular hollow, about 300 to 400 feet deep and rather more than a mile in diameter, containing at the bottom a shallow lake of salt water without any outlet, whose waters deposit crystals of sesquicarbonate of soda. The sides of the hollow to the north and northeast are absolutely level with the surrounding country, while in all other directions there is a raised rim, never exceeding 100 feet in height and frequently only 40 or 50, composed of blocks of basalt, irregularly piled, and precisely similar to the rock exposed on the sides of the hollow. The dip of the surrounding traps is always from the hollow, but very low.

"It is difficult to ascribe this hollow to any other cause than volcanic explosion, as no such excavation could be produced by any known form of aqueous denudation, and the raised rim of loose blocks around the edge appears to preclude the idea of a simple depression. It is true that there is no sign of any eruption having accompanied the formation of the crater ; no dyke can be traced in the surrounding rocks ; no lava or scoriae of later age than the Deccan trap period can be found in the neighborhood. The raised rim is very small, and cannot contain a thousandth part of the rock ejected from the crater, but it is impossible to say how much was reduced to fine powder and scattered to a distance, or removed by denudation.

"Assuming that this extraordinary hollow is due to volcanic explosions, the date of its origin still remains to be determined. That this is long posterior to the epoch of the Deccan traps is manifest, for the hollow appears to have been made in the present surface of the country, carved out by ages of denudation from the old lava flow. To all appearance the Lonâs lake crater is of comparatively recent origin, and if so it suggests that, in one isolated spot in India, a singularly violent explosive action must have taken place, unaccompanied by the eruption of melted rock. Nothing similar is known to occur elsewhere in the Indian Peninsula."

Besides the obstructions to drainage produced by overflows of lava, and by volcanic explosions, it may also be noted that volcanic dust and ashes, ejected from volcanoes during times of violent eruptions, may be deposited over the adjacent country in such a manner as to choke the streams and possibly form dams which would retain lakes. This process

[1] "A Manual of the Geology of India," 2d ed. Calcutta, 1893, pp. 19, 20.

has already been referred to in connection with the formation of basins through the action of eolian agencies.

Lava streams frequently cool on the surface while the liquid rock below is still flowing. In such instances, when the crust is sufficiently strong to sustain itself, the molten lava beneath flows out, leaving caverns. Openings of this nature may become water-filled and form subterranean lakes, or their roofs may fall in, leaving depressions open to the sky. Lakes and ponds occupying such depressions are thought to exist on the vast lava sheets of Oregon, Washington and Idaho, but clear, simple examples of the type are not at hand.

On a small lava flow on an island in Mono lake, California, there are depressions occupied in part by water, which are due to a general subsidence of the surface on account of the outflow of molten rock below and the crumpling of the crust into concentric, crescent-shaped ridges.

Another mode in which volcanic agencies may produce depressions is by subsidence of the surface about volcanoes, due to the removal of lava from subterranean reservoirs, but no instances where this has certainly occurred have yet been observed in this country.

Basins due to the impact of meteors. — The study of the origin of the crater-like forms on the surface of the moon recently made by Gilbert,[1] was suggested by the hypothesis that depressions on the earth's surface might result from the impact of meteoric bodies. This suggestion has already been referred to in describing Coon butte, and is one of great interest. Up to the present time, however, no basins on the earth's surface are known which can be ascribed to this agency.

If the earth was formed by the coming together of a large number of previously independent meteoric bodies, as is thought to have been the case by Lockyer, all evidence of such an occurrence in the relief of its surface is wanting. Small meteors are known to reach the earth every day, and a number have been discovered weighing many tons, but such an event as the earth coming in contact with a planetary mass a mile or several miles in diameter, as seems to have happened in the case of the moon, is not only unrecorded in history, but, as just stated, there is no evidence in the surface features of the earth to show that such an event has happened in recent geological time. If the earth once had a pitted surface, like the moon, and was scarred by vast crater-like

[1] " The Moon's Face," Philosophical Society of Washington, Bull. vol. 12, 1893, pp. 241-292.

depressions, each one the record of the piercing of its surface and the burial within its crust of a planetary mass previously revolving independently in space, the date of the last of the catastrophes which produced that condition must have been so remote that erosion has removed all surface evidence of the fact. Still farther negative evidence may be cited, inasmuch as no buried meteoric masses recognized as such, have been found in the rocks now forming the earth's surface. This is not proof, however, that the meteoric hypothesis, as applied to the earth, is not true, as the main events in that drama are assumed to have been enacted before the formation of the stratified rocks now recognizable.

Basins due to earthquakes. — During earthquakes there are undulations of the surface of the regions affected which sometimes produce permanent elevations and depressions and thus affect the drainage. The passage of earthquake waves through loose deposits may cause them to become more compact and perhaps produce depressions on their surfaces. In these and probably other ways, basins may be formed by earthquakes and give origin to lakes.

The best examples of lake basins in America, resulting directly from earthquake shocks, occur in what is known as the "Sunk country" in southeastern Missouri and northeastern Arkansas. A series of severe disturbances, known as the New Madrid earthquake, affected that region between 1811 and 1813, and caused both elevations and depressions in the forest-covered flood plain of the Mississippi. This region has recently been examined by W J McGee,[1] who reports that a low dome some 20 miles in diameter, was upheaved athwart the course of the Mississippi and that the river was held in check for a brief period, but soon cut a channel through the obstruction. An adjacent area some one hundred square miles in area, was depressed and is still, in part, occupied by lakes in which the trunks of trees killed by the inundation are standing.

During earthquakes in regions occupied by unconsolidated rocks, water is sometimes forced to the surface with great violence, probably on account of the compression of porous, water-charged strata, and rises fountain-like above the surface. The water brings with it quantities of sand and mud which are deposited around the points of discharge and serve to enlarge the depressions produced by the violent outrush. When the fountains cease to play these small crater-like basins remain as ponds.

[1] "A Fossil Earthquake," in Geol. Soc. Am., Bull., vol. 4, pp. 411–415.

Basins due to organic agencies. — The study of coral reefs has shown that bodies of sea water are sometimes cut off from the ocean, although rarely completely separated, by the growth of reefs of living coral adjacent to coasts, or as atolls about isolated islands and "banks." Lakes of this nature perhaps occur at the south end of Florida, and on the West India islands, but no well defined instances have been described.

The formation of peat in temperate latitudes affords another illustration of the manner in which organic agencies lead to the formation of lakes. The growth of the moss known as *Sphagnum*, from which peat is largely formed, may obstruct sluggish drainage ; and its unequal growth in swampy areas leads to the formation of mounds with depressions in their summits. The best known illustration of this type is Drummond lake, in Virginia, but many smaller examples occur in other swampy areas. It has been suggested that the basins in peat swamps may have originated by the burning of the bogs during times of excessive drouth. That this might happen is evident, but no authentic case of such an occurrence is known to the writer.

On the vast tundras skirting the Arctic ocean in both the Old and the New World, there are vast numbers of ponds and lakes held in depressions in the frozen bogs, and surrounded by banks of moss and other vegetation. These water-bodies have probably originated in various ways, but in some instances their birth may be traced to the luxuriant growth of vegetation in spring and early summer about the borders of lingering snow banks. When the vegetation of the tundras awakens after its long winter sleep, its growth is surprisingly rapid, and the snow drifts that last longest are surrounded with luxuriant mosses and brilliant flowers. When such accumulations of snow finally melt, the vegetation on the areas they occupied is less in amount than on the surrounding surfaces. The tundra increases in depth by the partial decay and freezing of the lower portion of the vegetation forming its surface, and the greatest thickness of frozen soil occurs where the vegetation is most luxuriant. For these reasons, the places where snow banks form year after year, become depressed in reference to the general surface, and give origin to lakes.

In sub-Arctic regions, as on the Aleutian islands, mosses and herbaceous vegetation grow luxuriantly, and among the hills sometimes obstructs the drainage by reason of the formation of a deep peaty soil by its partial decay.

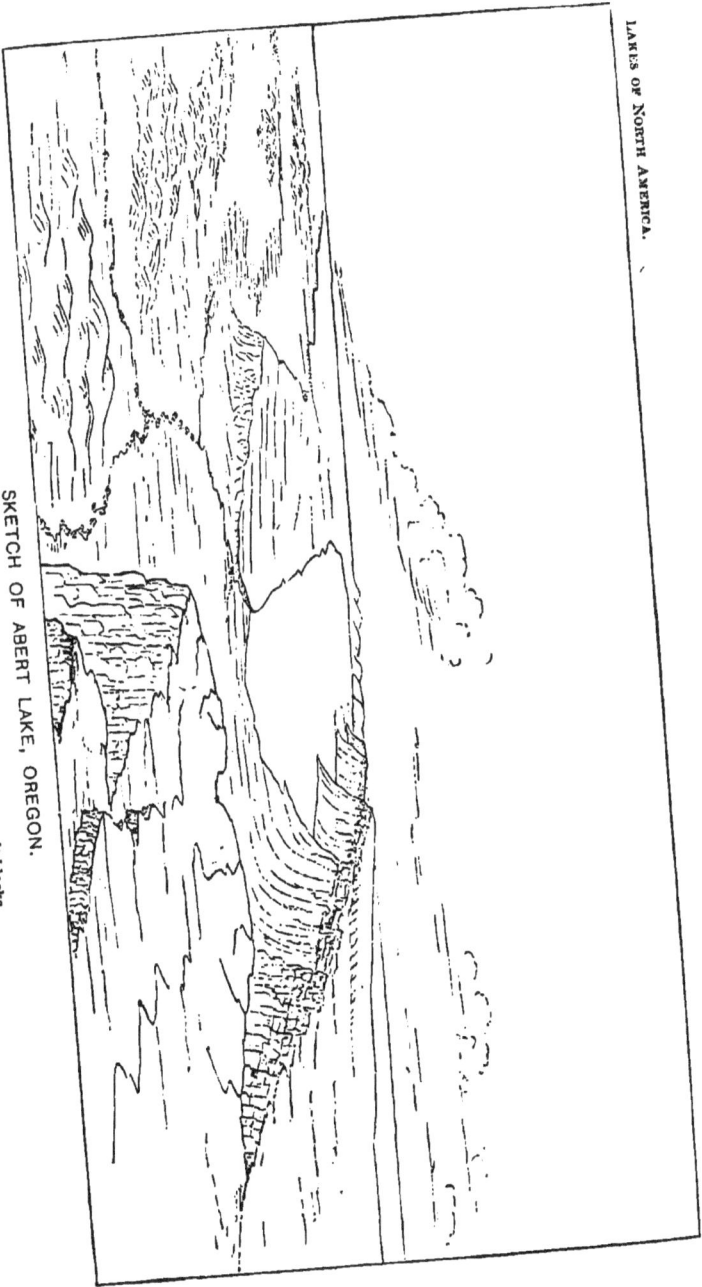

SKETCH OF ABERT LAKE, OREGON.

The cliffs are formed of the upturned edges of fault-blocks.

PLATE 6.

Beaver dams afford still another illustration of the manner in which drainage is obstructed and lakes formed by organic agencies. Beavers formerly lived over nearly the whole of North America, and are still found in limited numbers in the Northern states and Canada, and extending southward along the Cordilleras at least as far as New Mexico. The dams they constructed with great intelligence and skill, across small streams, retained drift logs and floating leaves, thus leading to the accumulation of deposits which obstructed the drainage for a long time after they had been abandoned by the animals that built them. The ponds and swamps due to the work of beavers number tens of thousands, and have produced important changes in the minor features of the surface of the continent. Many of these ponds, after becoming choked with vegetation and converted into peat swamps, have been drained and furnish rich garden-lands.

Where brooks and creeks flow through forested regions, it frequently happens that large trees fall across them and retain the sticks and leaves swept along by the current. When such a start is made, the mud carried, especially during freshets, is lodged among the leaves and branches, and tends still farther to obstruct the drainage and lead to the formation of swamps and lakes. This process has been observed especially in Red river, Louisiana, where timber rafts several square miles in area, and covered with living vegetation, form floating islands and dam the streams so as to cause their waters to spread out in shallow lakes twenty to thirty miles in length.[1]

Numerous instances in the Yukon river, in Alaska, were observed by the writer, where lateral branches of the stream and the passage ways between islands, were closed by accumulations of drift logs that greatly obstructed the flow of the waters. In some instances these accumulations, called "wood yards" by steamboat men, are several acres in extent.

Still another way in which organic agencies lead to the formation of basins may be observed in swamps where vegetable matter buried beneath mud and clay is undergoing decomposition. Openings similar to those produced in alluvial deposits by the violent escape of water during earthquakes, but not necessarily connected with such disturbances, are formed in marshes by the violent escape of gases from below. Instances of this occurrence have come under the writer's notice on Smoke Creek desert,

[1] Charles Lyell. "Principles of Geology." 11th ed., Vol. 1, p. 441. Humphreys & Abbott. "Report on the Physics and Hydraulics of the Mississippi," Professional Papers, Corps of Engineers, U. S. A., 1861, p. 37.

Nevada, and on swampy areas near Sandusky, Ohio. When these gas eruptions occur, the soft mud is sometimes thrown to a distance of one or two hundred feet, and conical depressions are formed which in some of the instances observed, are twenty feet or more in depth. The caving in of the banks' holes sometimes leads to the formation of pools fifty or sixty feet in diameter. The circular ponds frequently to be seen in swampy regions, when not due to encroaching vegetation, probably, in many instances, originated in the manner here noted.

The generation of gases, principally carbureted hydrogen, in the soft mud of the Mississippi delta, causes elevations known as "mud lumps," which in some instances are twenty-five feet high. Inequalities produced in this manner might easily lead to the obstruction of drainage and the formation of lakes, but no instance of such an occurrence seems to have been reported.

It has frequently been observed that cattle on visiting swampy places carry away considerable quantities of mud, adhering to their feet and matted in their hair. In arid countries where drinking places are usually small and widely scattered, they are visited by cattle and other animals in large numbers and a marked enlargement of the water holes is produced in the manner just stated. This process was more important when the plains of North America were densely inhabited by bisons. Many perennial pools and still more numerous depressions that are water-filled only during rainy seasons, are known as "buffalo-wallows," and are believed to owe their origin to a great extent to the carrying away of mud entangled in the thick hair of the animals after which they are named.

In the Appalachians there are several water holes, usually on the crests of ridges, that are called "bear-wallows," and are said to have been formed by bears that sought moist places in which to cool themselves during hot weather. As is well known, swine have a similar habit.

Basins due to movements in the earth's crust. — Great changes in the earth's crust have produced continents and ocean-basins, while smaller movements on land areas have resulted in the formation of mountains and valleys. During the growth of mountains it sometimes happens that the region between different systems or between two or more ranges, becomes enclosed so as to form a basin. This process has been in action in various localities since land first appeared, and during the course of geological eras must have resulted in the formation of many lakes ; but examples of water bodies of this type are rare at the present time, principally for the reason that the deformation of the earth's crust usually goes on slowly and the depressions formed are drained or filled with sediments as rapidly as they are formed.

The best examples on this continent of basins formed by the upheaval of mountains around them, occur in the great area of interior drainage between the Rocky mountains and the Sierra Nevada. The majority of the minor basins in this region, however, are due to secondary causes, but the vast seas, such as lakes Bonneville and Lahontan, which formerly existed there, occupied basins of the character here referred to.

The Laurentian lakes are held in basins produced in part by crustal movements affecting large areas, and in part by conditions resulting from other causes. Basins are also produced by less extensive elevations and depressions of the earth's crust. The corrugation of a region, owing to the formation of a series of approximately parallel folds, known as anti-clinals and synclinals, as in the case of the Appalachian mountains, must frequently produce basins in which water would be retained, were the process allowed to go on without some counteracting agency ; but here again, the movements are usually so slow that, especially in humid regions, the depressions produced are destroyed as rapidly as they are formed. While lakes in synclinal basins might be expected to be of common occurrence, they are in reality so rare that, so far as I am aware, none of the tens of thousands of the lakes of America can be pointed to as examples.

There is still another variety of earth movements in many instances less gradual than those referred to above, to which many lakes owe their origin.

Fractures in the earth's crust occur in disturbed regions and may be scores or even hundreds of miles in length. The edges of the broken strata on one side of a fracture are sometimes elevated, or those on the opposite side depressed, thus forming what is known as a "fault." The growth of faults sometimes goes on so slowly that no pronounced changes in topography result, for the reason that the rocks on the upheaved side of the fracture are eroded away as fast as they are raised. At other times, however, mountain ranges are produced, in which the strata are inclined away from the steep, broken face overlooking the line of fracture. In regions where such mountain ranges have been formed with comparative rapidity and where denuding agencies are weak, great disturbances in the drainage result, and "fault basins" are common. Numerous basins of this character occur in the Arid region and especially in Nevada and southeastern California, but probably the most typical example is the one occupied by Abert lake, Oregon.

Along the east side of Abert lake there is a long line of magnificent palisades, several hundred feet high, formed by the precipitous face of an eastward dipping fault block ; the lake washes the base of this escapement and occupies the depression formed by the subsidence of the rocks on the west side of the fracture. Something of the appearance of Abert lake, as seen from the crest of the palisades a few miles to south of its southern end, and also of the general structure of the underlying rocks, may be gathered from the accompanying illustration. The lake is about fifteen miles long with an average width of nearly four miles, and is shallow. It receives the water of a single creek, but does not overflow and is intensely alkaline.

Many of the lakes of the Arid region are of the Abert type, but usually the great depressions in which they occur have become deeply filled with the sediments of older water bodies, and they may be considered as occupying depressions on new land areas, or as belonging to the class of basins here considered, as one prefers.

In some instances the faulting that gave origin to the characteristic topography of the Great Basin region has been continued to the present time, and produced escarpments across the bottoms of the deeply filled valleys, so that the existing water-bodies are confined in part by recent fault scarps. An instance of this nature is furnished by Mono lake, California, which washes the base of a precipice formed by a recent movement of the great Sierra Nevada fault. A similar association has also been observed in connection with several of the lakes of western Nevada.

When a fault crosses the course of a river, the edge of the upturned block may rise so slowly that the stream is able to maintain its course and cut a channel through the obstruction as it is elevated, and a lake is not formed. Numerous instances of this nature have been observed by the writer in the central part of the state of Washington, where the Columbia and the Yackima river have eroded deep narrow gorges through the edges of fault blocks that were upheaved across their courses.

With basins produced by faulting, as in other instances of surface inequalities due to movements of the earth's crust, the question whether a lake will be formed or not, is answered mainly by the climatic conditions. In arid regions the surface effects of orographic movements are counteracted by erosion but slowly; while in countries with abundant drainage degradation goes on energetically, and unless the deformation of the surface is comparatively rapid, no pronounced topographic changes result. It is the ratio between the rate of deformation and denudation which determines whether basins shall be formed or not. Evidently the most favorable regions for studying the effects of movements in the earth's crust on the surface relief, are those in which the meteoric and aqueous agencies are least energetic, namely, in arid regions.

Basins due to land-slides. — On steep slopes great masses of rocks and earth not infrequently break away, especially after heavy rains, and descend suddenly as land-slides into the adjacent valley. When this occurs, the drainage in the valley may be obstructed so as to cause lakes to form. Avalanches of snow and loose rocks also produce similar results, but of a less pronounced character.

Small lakes originate in many cases on the surface of land-slides owing to the fact that such surfaces, after the descending mass has come to rest, usually incline toward the cliffs from which they broke away, in such a manner as to enclose basins. At times, a land-slide plows up the floor of the valley into which it plunges and forms a ridge, not unlike a terminal moraine, which may also act as a dam and hold a lake in check. Examples of basins formed in each of these several ways have been examined by the writer in the state of Washington [1] and in other regions, but need not be described at this time.

Basins due to chemical action. — In limestone countries the drainage is often subterranean and finds its way through caverns formed by the solution of the rock. The roofs of such caverns fall in as the general erosion of the region progresses, and obstruct the drainage channels so as to form lakes. The surface waters reach underground channels through openings termed "sink-holes," or "swallow-holes," which are enlarged by solution, and frequently become closed so as to hold ponds. In portions of Kentucky and throughout the Great Appalachian valley, where the underlying rock is limestone, circular ponds of this nature are so numerous that they give character to the landscape. Lakes also occur in the caverns themselves, owing to various causes, the most frequent being the falling of portions of their stalactic roofs, as may be seen in Mammoth and Luray caverns.

[1] "Geological Reconnoissance in Central Washington," U. S. Geol. Surv., Bulletin No. 108.

Basins of small size, due to chemical precipitation, occur in connection with springs that deposit calcareous tufa or siliceous cinter. Many examples of pools formed in this way occur in the Yellowstone National Park and in other hot spring regions of the Cordilleras. Near the west shore of Mono lake, California, there is a castle-like bowl of calcareous tufa, fully 50 feet high and from 150 to 200 feet in diameter, with several long aqueduct-like branches, which was formed from the water of a spring that has now ceased to flow. Far out on the desert valleys of Utah and Nevada one sometimes finds circular basins with rims of tufa from a few inches to three or four feet high, and holding beautifully clear water with a temperature approaching the boiling point. In other instances, these deposits rise several feet above the adjacent surface and resemble volcanic craters. In their summits there are frequently steaming caldrons.

In regions underlain by gypsum, rock salt, and other easily soluble substances, depressions are formed on account of the removal in solution of the rocks beneath and the subsidence of the surface.

Gypsum is thought by some geologists to owe its formation to the alteration of limestone by the passage through it of sulphurous gases or of sulphurous waters. When this occurs, the volume of the deposit is increased and the ground above may be elevated into mounds, and thus obstruct the drainage.

CONCLUSION.

In this chapter an attempt has been made to describe briefly the principal types of lake basins occurring in North America, to indicate the processes by which they have been formed, and to show to some extent, where they severally belong in the history of topographical development.

Many basins have resulted from the action of more than one agency, and in not a few instances several agencies have coöperated in their production. Basins of a composite character have thus originated, but the principal cause leading to their existence is usually so pronounced that when carefully studied, they may without great violence be referred to some one of the types here described.

The study of lakes has shown that they frequently have a long and varied history, which is no less interesting and instructive than the story of the origin and decadence of the hills that are reflected in their glassy depths. Some of the phases of their not uneventful lives are described in the succeeding chapters.

CHAPTER II.

MOVEMENTS OF LAKE WATERS AND THE GEOLOGICAL FUNCTIONS OF LAKES.

Tides. — The waters of lakes are influenced by the attraction of the sun and moon in the same manner as the waters of the ocean. Owing to the comparatively small extent of inland water-bodies, however, the rise of their waters is so small that it is not noticeable, and can only be determined by refined measurements.

Observations made by the U. S. Lake Survey at Chicago, have shown that Lake Michigan has a tide with an amplitude of $1\frac{1}{2}$ inches for the neap tide and 3 inches for the spring tide.

Waves and currents. — The waters of fresh lakes respond to the influences of the wind more quickly than the heavier waters of the ocean, but the waves produced are smaller and less regular than in the open sea. On the Laurentian lakes, waves from 15 to 18 feet in amplitude have been observed during long continued storms. The heavy ground swell of the ocean is but faintly reproduced by the fresh water "seas." During rough weather on the lakes the waves are more like the short, "chop seas" than the heavy surges of the open ocean.

The friction of the wind on the surfaces of lakes produces very decided movements in their waters. In their central portions, especially, there are frequently strong currents due to this cause, in addition to the slow movement of the waters toward an outlet. A study of the currents of the Laurentian lakes has been undertaken by the United States Weather Bureau, by means of bottles containing a record of the locality where they were set adrift and a request that the finder will note the locality where they are recovered and transmit the record to the Chief of the Weather Bureau. The results of observations made in the summer season of 1892 and 1893, have been published,[1] and the general courses of the currents so far as ascertained, indicated on a chart which is reproduced on Plate 7. The effects of the prevailing westerly winds on the surface movement of certain of the Laurentian lakes, is indicated by the trend

[1] U. S. Department of Agriculture, Weather Bureau, Bulletin B.

of the principal currents. When the larger axis of a lake coincides with the direction of the prevailing winds, a surface current is established through its center, as in the case of lakes Erie and Ontario, with return currents and eddies along the shore and about islands. When lakes lie athwart the prevailing winds the main currents combine with the return currents and form minor swirls, as is shown on the chart in the case of lakes Michigan and Huron. In Lake Superior there is a general circulation which follows the main shore lines, but its course has not been fully determined. It has been found that the currents of the Laurentian lakes have in general a speed of from 4 to 12 miles a day, but in certain observed instances, this is increased to $2\frac{1}{2}$ to 4 miles an hour or from 36 to 96 miles a day.

The currents in the central part of a lake produce slight if any changes on the topography of its basin, but when they follow the shore important results may follow. When the wind blows obliquely to the shore, strong currents are frequently produced which follow the general trend of the coast, but cut across bays and inlets. These currents, with the assistance of waves, sweep along sand and gravel, and produce important changes on the bottom, particularly when the water is shallow. The rôle played by waves and currents in modifying topography is considered with some detail in the next succeeding chapter.

Strong winds blowing in a nearly uniform direction for several days cause the waters of lakes to move with them, and to rise on the shores against which they are driven, so as frequently to produce disastrous inundations. A gale blowing from the north over Lake Michigan has been observed to cause a rise of seven feet at Chicago. In November, 1892, a storm from the west caused the waters of Lake Erie near Toledo, to fall between eight and nine feet below the normal fair weather level. At the same time, unusually high water was experienced at the east end of the lake. The differences in the level of the waters of Lake Erie, at Buffalo, between a high-water stage produced by an eastward blowing gale, and a low-water stage accompanying a westward or off-shore gale, has been observed to amount to $15\frac{1}{2}$ feet. An eastward movement of the waters of Lake Superior has been known to accompany a gale from the west and to produce an unusual rise in the water of St. Mary's river.

The height to which the waters reach on lake shores, owing to strong winds, establishes the upper limit of wave-action, and leads to the formation of storm beaches at an elevation of several feet above the normal

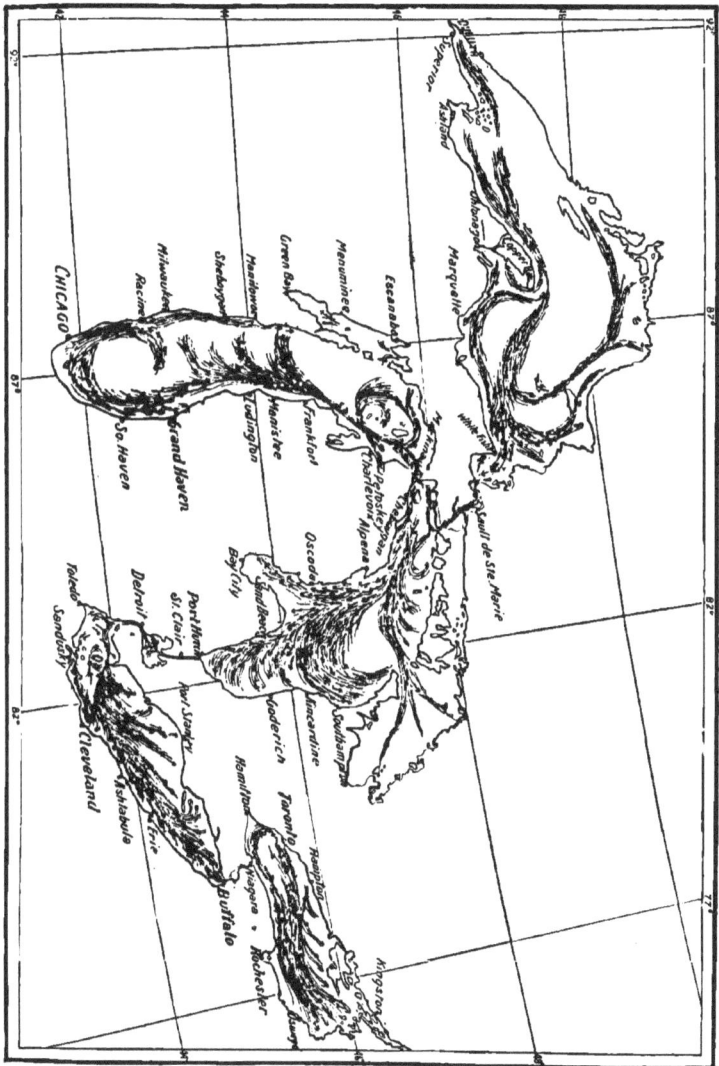

PREVAILING CURRENTS IN LAURENTIAN LAKES. (AFTER U. S. WEATHER BUREAU.)

PLATE 7.

stage. When the shores of a lake are low, broad areas are inundated during storms that sweep the waters towards them. New outlets may be established at such times across low divides, and lead to important changes in drainage.

Seiche. — Lake waters are also sensitive to changes in atmospheric pressure. In some instances variations of level during calm weather, amounting to several feet, have been observed, and are supposed to be due to sudden changes in barometric pressure on different portions of the water surface. Besides these larger movements, which can be correlated with atmospheric changes, and are known as *seiches*,[1] there are certain rhythmical pulsations producing a difference of level of as much as four or five inches during calms, when no variation in atmospheric pressure of an analogous character can be detected. These minor movements are not thoroughly understood.

These and other changes of a similar nature are of great interest in connection with meteorological studies, but have little if any geological significance.

It is to be expected that earthquakes would produce "tidal waves" in lakes similar to those occurring in the ocean, but observations in this connection are wanting.

Temperature. — Lake waters are warmed by the sun's rays and by contact with the air. It has also been thought by some that very deep lakes may have their bottom temperatures modified by the general internal heat of the earth, but observations do not seem to support this conclusion. Water is a poor radiator and an indifferent conductor of heat, and does not respond to atmospheric changes of temperature as quickly as do rock surfaces. Shallow lakes are warmed throughout by the summer's heat and chilled to the bottom by the winter's cold ; but their temperature is much more uniform than that of the adjacent air. The shallow lakes of the Northern states have been found to have a nearly uniform temperature during the summer months of 75° Fahrenheit. In winter their temperature in general is, of course, 32° Fahrenheit. In lakes having a depth in excess of about 800 feet, more interesting conditions are found. The temperatures of deep lakes are ascertained by means of self-registering thermometers attached to sounding lines. In this

[1] E. A. Perkins. "The Seiche in American Lakes," American Meteorological Jour., Oct., 1893.

way accurate measurements of temperature at various depths have been
made in a number of lakes, both in America and in Europe, with remarkably consistent results. Of the observations thus far made in this country,
the most instructive are by Professor John Le Conte,[1] in Lake Tahoe,
California. From the report of these observations I quote the following :

"These experiments were executed between the 11th and 18th of August, 1873.
The same general results were obtained in all parts of the lake. The following table
contains an abstract of the average results, after correcting the thermometric indications
by comparison with a standard thermometer :

| Obs. | Depth in Feet. | Depth in Meters. | Temperature. | |
			Fahrenheit Scale.	Centigrade Scale.
1	0 = Surface.	0 = Surface.	67°	19.44°
2	50	15.24	63°	17.22°
3	100	30.48	55°	12.78°
4	150	45.72	50°	10°
5	200	60.06	48°	8.89°
6	250	76.20	47°	8.38°
7	300	91.44	46°	7.78°
8	330 (Bottom)	100.58	45.5°	7.50°
9	400	121.92	45°	7.72°
10	480 (Bottom)	146.30	44.5°	6.94°
11	500	152.40	44°	6.67°
12	600	182.88	43°	6.11°
13	772 (Bottom)	235.30	41°	5°
14	1506 (Bottom)	459.02	39.2°	4°

"It will be seen from the foregoing numbers that the temperature of the water
decreases with increasing depth to about 700 or 800 feet (213 or 244 meters), and
below this depth it remains sensibly the same down to 1,500 feet (459 meters). This
constant temperature which prevails at all depths below say 250 meters is about
4° C. (39.2° Fahr.). This is precisely what might have been expected; for it is a
well-established physical property of fresh water, that it attains its maximum density at
the above-indicated temperature. In other words, a mass of fresh water at the temperature of 4° C. has a greater weight under a given volume (that is, a cubic inch unit
of it is heavier at this temperature) than it is at any temperature either higher or
lower. Hence, when the ice-cold water of the snow-fed streams of spring and summer
reaches the lake, it naturally tends to sink as soon as its temperature rises to 4° C.;
and, conversely, when winter sets in, as soon as the summer-heated surface-water is

[1] "Physical Studies of Lake Tahoe," Overland Monthly, 2d Series, vol. 2, 1883, pp.
506–516, 595–612; vol. 3, 1894, pp. 41–46.

cooled to 4°, it tends to sink. Any further rise of the temperature of the surface-water during the warm season, or fall of temperature during the cold season, alike produces expansion, and thus causes it to float on the heavier water below; so that water at 4° C. perpetually remains at the bottom, while the varying temperature of the seasons and the penetration of the solar heat only influence a surface stratum of about 250 meters in thickness. It is evident that the continual outflow of water from its shallow outlet cannot disturb the mass of liquid occupying the deeper portions of the lake. It thus results that the temperature of the surface-stratum of such bodies of fresh water for a certain depth fluctuates with the climate and with the seasons; but at the bottom of deep lakes it undergoes little or no change throughout the year, and approaches to that which corresponds to the maximum density of fresh water."

Influence of lakes on climate. — Inland water bodies exert an important influence on the climate of their shores, in reference especially to temperature and humidity, and also on the direction and character of the more gentle winds. The surface waters of lakes receive their temperature in a great measure from the air in contact with them, and are warmed or cooled at rates having some relation to their depth. The temperature of shallow lakes varies but little from that of the adjacent atmosphere, but changes less rapidly for the reason, already stated, that water surfaces are poor radiators. The differences between the rates of radiation between adjacent land and water surfaces, affect the temperature of the air above them, and in calm weather give origin to lake and land breezes.

The tens of thousands of small lakes scattered over the glaciated portions of North America have an important combined influence on the general climate, although their effects may, perhaps, be difficult of direct determination. These lakes cool and moisten the atmosphere by evaporation during the hot summer months, and when they freeze as winter approaches, a vast amount of "latent heat" is liberated and moderates the fall in temperature. It is stated by physicists that every ton of water converted into ice gives out as much heat as would be required to raise the same quantity of water from 30° to 174° Fahrenheit. A reverse process, not so congenial to the welfare of man, takes place, however, when the ice melts in spring, as then an amount of energy equal to that previously lost, must be again absorbed in order that the ice may change to water. The warm southern winds are thus chilled and the opening of the flowers delayed.

Deep lakes, as already seen, have a uniform bottom temperature of 39 degrees of the Fahrenheit scale, and do not freeze in winter except about their shores where the water is shallow, for the reason that the low temperature of the air above them does not continue long enough for the entire water-body to become cooled to the degree of maximum density. Until this happens the water cooled at the surface from contact with the air, has its density increased and sinks, and is replaced by warmer and consequently lighter water, rising from below, and ice cannot form.

The surface waters of deep lakes are thus above the mean temperature of the adjacent atmosphere in winter; but in summer they are cooler than the air, as the warmed surface layer loses heat by conduction downward. The winds that blow over them are thus tempered in a manner congenial to the growth of vegetation both in warm and in cold weather.

The influence of the Laurentian lakes on the climate of their shores is well marked, as was clearly shown many years since by Alexander Winchell.[1] On charts that have appeared showing the winter and summer isobar, that is, lines drawn through the various localities having the same mean temperature, the lines showing the mean summer temperature curve northward in the vicinity of Lake Michigan especially, while the lines indicating mean winter temperature present an equally marked southern curvature, showing that the lakes cool the air that passes over them in summer and warm it in winter. The genial influence of the lakes is also plainly to be seen in the distribution of the fruit-belts of Michigan, Ohio, and New York.

If we should construct a map showing the mean humidity of the air, by drawing lines through the localities having the same "relative humidity," the influence of the lakes would be quite as apparent as in the case of the isothermal lines, but the curvature in both winter and summer would be southward.

The amelioration of climate produced by large inland water-bodies has an important influence on the flora and fauna of their borders, and therefore on the character of the fossils entombed in their sediments. Another fact of geological interest in this connection, is that rocks decay more rapidly under warm, moist climates than in arid or in Arctic regions, and deeper and richer soils are produced. This, again, influences the life of lake regions, and is, perhaps, of sufficient importance to be considered in interpreting the records of ancient inland water-bodies.

Influence of lakes on the flow of streams. — Lakes act as storage reservoirs and regulate the flow of the streams, of which they are enlargements. In the case of a river subject to sudden freshets, the disastrous

---

[1] "The isothermals of the Lake Region," Am. Assoc. Adv. Sci., Proc., vol. 16, Troy Meeting, 1870, pp. 106–117.

effects of a sudden rise would be checked, and even entirely averted, if a lake of sufficient size existed in its middle course, or if there were a number of lakes on its tributary streams.

The modulating influence of lakes on the flow of streams is well known to hydraulic engineers; and it has been proposed to regulate the flow of the Mississippi by building storage reservoirs on its head waters. Such reservoirs could be filled during floods and the water allowed to escape when the danger stage had passed. In this manner the disasters resulting from annual freshets could be averted and navigation improved during the seasons of low water.

The effect of gales in heaping up the waters of lakes on the shores against which they blow has already been noted, and an instance cited where the waters of St. Mary's river were suddenly raised by a gale on Lake Superior. A rise of the water in streams flowing from large lakes, due to this cause, is exceptional, however, and by no means as destructive as the fluctuations produced by storms and melting snow on water courses that are without the regulating influences of large lakes.

The sudden escape of lakes held by dams of ice also causes floods in the streams below, as in the case already cited of the Rhone, when Märjelen lake is drained, and of the Stickeen, where the glacial-held lakes on its tributaries break their icy bands.

The rise of Lake Bonneville until it found an outlet and then rapidly cut down its channel of discharge through unconsolidated material, as will be described in advance, is supposed to have caused a great rise in Snake river, to which it became tributary. In these and other ways that might be cited, it appears that lakes may cause floods in their draining streams as well as avert them.

Lakes as settling basins. — The streams flowing into lakes are frequently turbid and heavy with sediment, especially after storms, but the rivers flowing from them are usually clear and free from all but possibly the finest of material in suspension. During the slow passage of the waters through a lake which has an outlet, the material in suspension falls to the bottom and contributes to the filling of the basin, while the clarified waters flow on.

The fact that bodies of standing water retain the mineral matter brought to them in suspension, is illustrated more or less perfectly in nearly every lake and pond, and even by ephemeral pools by the wayside, but is especially marked in great seas like those drained by the St. Lawrence. During storms, all of the streams pouring into the upper Laurentian lakes, from the surface drainage of the land, are brown and heavy with mud, but the water rushing over Niagara remains of the same deep greenish-blue tint season after season and year after year. Niagara river, above the falls, and the St. Lawrence are surface streams, because their clear waters have but slight power of corrosion; it is for this reason that during the centuries they have occupied their present channels they have not materially deepened them.

In the case of lakes fed by the turbid waters from glaciers, the part they play as settling basins is even more strikingly shown than in the instances just cited. Lake Geneva, Switzerland, fed by the silt-laden waters of the Rhone, is discolored for several miles from where the river enters, but when the waters leave the lake and again start on their journey they are wonderfully clear. An abundance of similar illustrations are furnished by the glacial-fed lakes of the Sierra Nevada and Cascade mountains and by some of the numerous lakes on the head waters of the Yukon.

The streams flowing from lakes are not always clear, however, as exceptions occur where the outlets are so situated that shore currents may bring sediment to them. The construction of beaches and embankments by shore currents may take place at the outlet of a lake so as to obstruct the escape of its waters and initiate a struggle between the waters tending to form deposits and those escaping through the channel of discharge. The outflowing waters may thus be rendered turbid and have the material supplied with which to erode their channels. A case in point is thought to be furnished at the south end of Lake Huron, where River St. Clair has its source, although definite observations on the relation of the outlet to shore currents have not been made. The waters of River St. Clair are not of the transparent character that would be expected in a stream starting from a large lake; and a broad delta has been formed in Lake St. Clair, into which the river empties after a short course through low alluvial lands. The source of the material forming the delta cannot be referred to the erosion of the banks of the stream, and is not furnished by tributaries, but apparently comes from the action of waves and currents on the shores of Lake Huron adjacent to its outlet.[1]

The rapidity with which lake basins in all parts of the world are becoming filled with sediment is sufficient in itself to show that no lakes fed by turbid streams can be geologically old.

[1] These conclusions have recently been confirmed by F. B. Taylor in an instructive paper on " The Second Lake Algonquin." Am. Geol., vol. 15, March, 1895, pp. 171, 172.

Mechanical sediments. — The coarse sediment brought to lakes by streams is either built into deltas or swept along the coast by shore currents and mingled with the pebbles and sand derived from the wear of the land by shore waves. The finer products of the wash of the land, and of shore erosion, are carried lakeward and deposited in stratified layers over the lake bottom. In general, the sheet of material thus spread out is thickest and coarsest near shore and becomes finer and thinner as the distance from land increases. When sedimentation goes on uninterruptedly until a basin is filled, the result is a more or less regular lens-shaped body of sediments, having a broad central area of fine material, which graduates into a fringe of coarser character about its borders. The coarse strata in the shore deposits overlap and dovetail along their lakeward margins, with the outer borders of the layers of fine sediment in the central part of the basin, for the reason that the coarser material is carried farther from land during storms than when the weather is calm. This general relation of coarse shore and fine off-shore deposits is of interest, especially in the study of extinct lakes, and may enable one to draw their former boundaries with considerable accuracy even when all distinctive features of their shore topography have been obliterated.

The sediments of the existing lakes of America, so far as they have been studied, are principally clays, which vary in character according to the nature of the rocks and soils on the neighboring land. The sediments of the Laurentian lakes and of lakes generally, particularly in humid regions, are characteristically blue clays. The Pleistocene clays of the Erie and Ontario basins are tenacious blue clays, similar to those now accumulating in the same basins; but the clays deposited during a former broad extension of Lake Superior are fine, evenly laminated pinkish clays, and owe their distinctive tint to the color of the rocks from which they were derived.

The sediments now accumulating in the lakes of the arid regions, but more especially in the temporary or playa lakes, are usually light-colored, and have a yellowish tint when dry.

In regions of deep rock decay, like the southern Appalachians, the débris swept into lakes would have the characteristic tints of *terra rossa*, as the highly oxidized product of prolonged rock decay is termed, unless it was mingled with organic matter in sufficient quantity to deoxidize the iron to which its richness of color is due.

The generalization that all lake sediments are of a reddish tint, formerly advanced by certain English geologists, does not find support from somewhat extended observations made in this connection in America. In fact, the blue and yellowish tints of such deposits are so general in this country that the reverse of the proposition referred to might be more reasonably claimed.

In small lakes, when sedimentation is retarded, the growth of mollusks, diatoms, etc., may progress rapidly and their dead shells accumulate on the bottom so as to exceed the amount of mechanical sediment, and shell marl and diatomaceous earth be formed. This process is especially well marked in lakes that are surrounded by matted vegetation through which the inflowing waters percolate and are filtered of nearly all material in suspension. As the growing mosses encroach on lakes of this character, a layer of peat is formed above the marl and a well-marked stratification results. Layers of peat above strata of shell marl may be seen in process of accumulation in many of the small lakes of Michigan and other similar regions. In lake and swamp deposits that are now drained and utilized for farming purposes, a layer of white marl beneath black humus, is frequently exposed. These deposits have an additional interest from the fact that we find in them the bones of the mastodon, mammoth, giant beaver and huge sloth-like animals that roamed over North America in recent times, but are now extinct.

PLATE 8.

SEA-CLIFF IN BOULDER CLAY, WITH BEACH IN FOREGROUND. SOUTH MANITOU ISLAND, LAKE MICHIGAN.

CHAPTER III.

THE TOPOGRAPHY OF LAKE SHORES.

THE variety and beauty of a landscape, embracing mountains and hills, valleys and ravines, is mainly due, as is well known, to the action of running water. The lines resulting from this mode of sculpture are more or less vertical. The waters of lakes also engrave their histories on the rocks, but the writing conforms with the water surface and is in horizontal bands. Two strongly contrasted types of relief are thus produced, which may be distinguished at a glance. The details in each type may be separated and their mode of origin explained. Each feature of the land is thus found to have a meaning, and the pleasure derived from even the most sublime and beautiful landscapes is vastly enhanced to those who can read their histories.

The work of rain and rivers is outside the scope of the present book, but the principal topographic features characteristic of lake shores will be briefly described and their mode of origin indicated.

The sea cliff. — Usually the first features of a lake shore to attract attention are the steep slopes which rise from the water's edge and seem to mark the boundary beyond which the waves cannot pass. That the slopes here referred to have been produced by the waters of the lake eating into the land, is so apparent that it seems almost a waste of words to explain the process by which they are formed. Their declivity varies according to the nature of the material forming the land and also in conformity with atmospheric conditions. When the shores are of soft rock or loose unconsolidated material, the slopes are gentle, but when the shore is of hard rock they may become vertical or even overhanging precipices. In regions where weathering is progressing actively, the waste of the land, owing to the combined influences of rain, frost, etc., may be more rapid than the erosion of a lake shore by waves and currents; under these conditions the bluffs bordering a lake will have a more gentle slope than where atmospheric agencies are relatively less destructive. The name " sea cliff " is applied to the slopes produced by the under-cutting of lake shores without reference to their declivity, and has been borrowed from the nomenclature of the oceanic shores where topographic forms similar in character and in origin exist in many places on a magnificent scale. Variations in the appearances of sea cliffs in soft and hard material are shown on Plates 8 and 9. These illustrations have been selected from a large number of photographs taken by the writer on the borders of the Laurentian lakes, and illustrate the two types of shore features there most pronounced.

The recession of sea cliffs may be best studied when a gale is blowing directly on shore. At such a time, each wave as it reaches shallow water and surges up on the land, carries forward the gravel and sand within reach and dashes it against the base of the cliff and tends to wear it away. The finer products produced by the friction and pounding of the loose stones against each other and against the cliff, are carried lakeward by the under-tow, leaving the coarser fragments ready to be caught up by the next inrush of water and the process repeated. As the cliff is under-cut, fresh angular fragments fall from its face to the beach below and are at once attacked by the waves and sooner or later reduced to rounded gravel and sand. The cliff thus furnishes the tools for its own destruction.

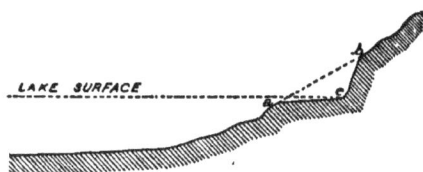

FIG. 2.— PROFILE OF A SEA CLIFF AND TERRACE.

The manner in which lakes wear away the land confining them is illustrated in the following section of a rocky shore, which also shows the relation of the sea cliff *b c* to the platform or terrace *a c* at its base.

Waves are only able to reach the land in a narrow vertical interval, determined mainly by the difference in their height during calm weather and when storms are raging. Even in the case of large lakes this interval does not exceed ten or fifteen feet, and on account of the débris usually encumbering the shore, the actual zone of erosion on the fresh rock surface is normally very much less than this. The waves thus act like a horizontal saw cutting into the land. The result is that at the base of every sea cliff there is a platform or terrace, as indicated in the above diagram. The junction of the sea cliff with its accompanying terrace is a horizontal line, determined by the elevation of the lake surface.

Lake waters unaided by débris, like the waters of clear streams, have but slight power to erode. It is only when the margin of a lake is sufficiently shallow to bring the débris on its bottom within the reach of the

SEA-CLIFF IN HARD SANDSTONE, WITH BEACH BEYOND. AU TRAIN ISLAND, LAKE SUPERIOR.

PLATE 9.

waves that the land is cut away so as to form sea cliffs and terraces. This is shown in a striking manner along large portions of the shores of Lake Superior, where bold cliffs, an inheritance from a previous topographic cycle, plunge into deep water, and are without talus slopes or other loose deposits within reach of the waves. In these instances there is scarcely a mark on the rocks that would record the present horizon of the lake should its waters be withdrawn. Clear waters may dissolve the rocks against which they dash, however, and when cliffs of limestone and other easily soluble rock descend into deep waters, a line of grottoes and caves may be formed below the upper wave limit, and perhaps increase until a shelf is produced on which sand and pebbles could lodge. When this happens, erosion by solution is assisted by mechanical means, slight at first, but increasing as the conditions become more favorable, until cliffs and terraces result.

Terraces. — The terraces about the margins of existing lakes are usually covered with the loose stones and sand, and form the beaches on which one may walk during calm weather.

The surface of a typical lake terrace slopes gently lakeward and is bounded on the landward margin by the upward slope of the accompanying sea cliff, and on the submerged, lakeward margin by a downward slope leading to deeper water. These terraces owe their formation to excavation or to deposition, and in most instances the two processes are combined. Even when the terrace is due principally to excavation, there is a surface layer of rounded débris resting on it, which is usually thickest on the lakeward margin and forms the lakeward slope. These features are shown in the following cross section of a lake shore, where a compound terrace is being formed, and also on Plate 13.

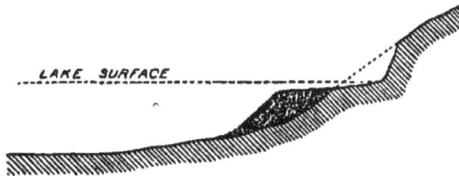

FIG. 3. — PROFILE OF A CUT AND BUILT TERRACE.

On precipitous, rocky shores, terraces are not produced, for the reason already stated in considering the origin of sea cliffs, that the débris from the land falls into deep water below the reach of the waves.

Reference has already been made to the action of the waves when the wind blows directly on shore. The return current is then an undertow flowing lakeward. When the wind blows against the shore at a low angle, however, currents are established which travel along the lake margin and sweep the loose material on the surface of the terrace with them. These currents have many of the features of streams, and greatly increase the power of waves to erode the land. The upward movement of waves tends to lift loose material within their reach and the lateral movement of currents to transport it. The loose material at the base of sea cliffs is thus carried along the beach by shore currents in one direction or another, according to the direction of the wind, and deposited so as to form accumulations of various character.

When a headland, with a beach at its base, is flanked on either hand by low shores, the débris falling from its face is carried along by the shore currents and built into terraces adjacent to the land or deposited so as to form free embankments or ridges, at some distance from the original shore. That this process is of common occurrence may be shown on many lake margins by examining the material forming rocky headlands and comparing it with the stones on neighboring beaches. In such instances the rock fragments at the base of the cliff will frequently be found to be large and angular and to become smoother and more and more rounded the farther they are traced from their parent ledges.

Terraces and marginal embankments, built wholly of gravel and sand, may also be formed on low shores by the washing up of loose material from this lakeward margin, thus deepening the water on the outside of the shelf.

The transportation of débris along the surfaces of terraces by the combined action of waves and currents, and its deposition when deep water is reached, leads to the formation of structures of various forms, known as embankments.

Embankments. — This name has been adopted for free ridges of loose material built by currents about the margins of water-bodies. They have the general form of railroad embankments, and their level crests in most instances rise from a few inches to, perhaps, three or four feet above the calm-weather surfaces of the water in which they occur. The tendency of built terraces to change to embankments on low shores has already been noticed, but the most typical examples occur where shore currents, having an abundance of loose material at their command, are deflected

SHORE OF LAKE BONNEVILLE, WELLSVILLE, UTAH. (AFTER GILBERT.)

An Illustration of the contrast between Littoral and Subaërial Topography.

PLATE 10.

into deep water and thus lose their power to transport. The variations in the shapes of embankments have led to the recognition of various more or less specific forms, such as spits, loops, bars, V-bars, etc., some of which are described below.

The building of embankments can be best studied where there is an abrupt change in the direction of the shore adjacent to a locality where the formation of a sea cliff and its accompanying terrace is in progress. Such an instance is illustrated in the following sketch-map :

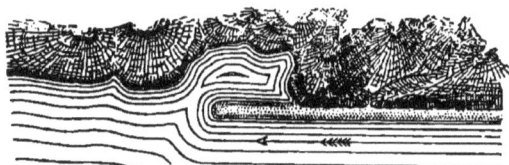

FIG. 4. — SKETCH-MAP OF AN EMBANKMENT.

The shore on the right of the cove is steep and forms a sea cliff that rises above a terrace along which the current travels in the direction indicated by an arrow. Shore currents follow the broader outlines of the land, but cut across bays and inlets. For this reason, in the case before us, the sand and gravel swept along the surface of the terrace is carried into deep waters and is deposited when the direction of the shore changes abruptly, as the flow of the water is then checked. The terrace is prolonged as an embankment, having the same level, and is lengthened by material carried along its surface and deposited at its distal extremity. The construction of such an embankment is analogous to the manner in which railroad embankments are made by carting dirt along them from a cut and dumping it at the end of the unfinished structure. In cross sections an embankment shows a more or less perfect arching of the material, and forming what may be termed an "anticlinal of deposition."

In the ideal illustration here presented, it is evident that a continuation of the process would result in the prolongation of the embankment until it touched the shore at the left of the bay. The outline of the lake would then be simplified and a lagoon formed behind the embankment. Should a stream enter such a lagoon, the water escaping from it might keep a channel open to the lake, but a struggle would ensue between the shore currents tending to close the break and the outflowing water striving to keep it open. Eddies in the conflicting currents would result and lead to changes in the outlines of the embankment.

When a structure like that described above is incomplete and projects from the shore like an unfinished railroad embankment, it is termed a *spit*. An illustration of such an instance observed on the shore of Au Train island, Lake Superior, is shown in Plate 11.   See also Plates 2, 3 and 4.

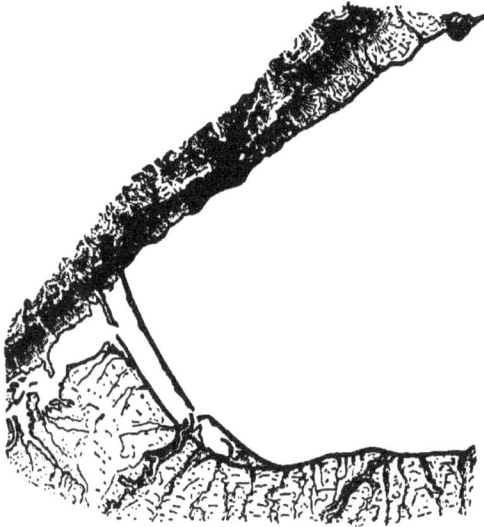

When an embankment spans the entrance of a bay so as to shut it off more or less completely from the main water body, it is termed a *bar*, in accordance with the custom of mariners in designating such obstructions to navigation.   Maps of bars on the shores of lakes Superior and Ontario are reproduced in Figs. 5 and 6, from the maps of the U. S. Lake Survey.   The manner in which these were formed, as well as their various modifications

FIG. 5. — MAP OF SAND BARS: WEST END OF LAKE SUPERIOR.

of outline and the presence of channels across them in certain instances, will be understood from the description of a more simple example just given.

The end of a spit is frequently turned toward the shore, owing to a deflection of the current that built it, or to the opposing action of two or more currents, and becomes a *hook*, as is illustrated on Plate 12. Again, where the hook is more pronounced and the distal end of the structure touches the shore, as happens occasionally when there are only slight changes in the direction of the coast line, a *loop-bar* or *V-bar* results.

In brief, it may be said that the waves and currents of lakes have the power of excavating cut terraces along the shores confining them and of carrying away the waste from the cutting, together with similar material contributed by streams, and of building it into terraces and embankments of various forms adjacent to neighboring shores.

Deltas. — Where streams bring to a lake more detritus than is carried away by shore currents, accumulation takes place and an addition, termed

SPIT OF SHINGLE, AU TRAIN ISLAND, LAKE SUPERIOR.

PLATE II.

a delta, is made to the land. The most instructive deposits of this nature occur where high grade streams enter a lake, as when a lake washes the base of a mountain range. In such an instance, pebbles and water-worn boulders are swept along by the stream until it mingles with the quiet lake water, where its velocity is checked and the coarser portion of its

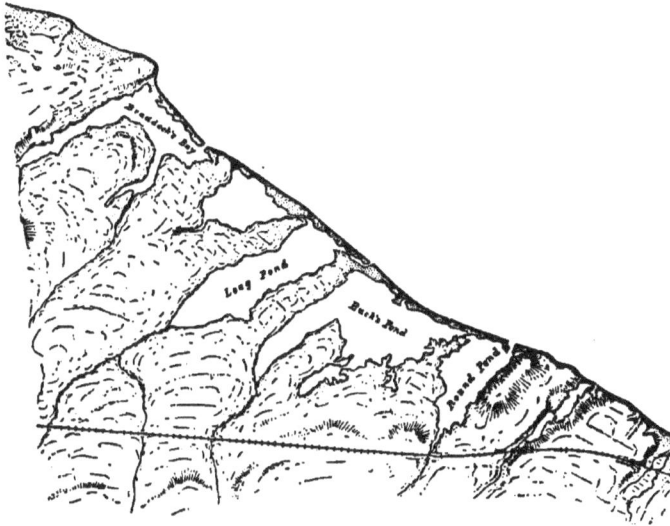

FIG. 6. — MAP OF SAND BARS: SOUTH SHORE OF LAKE ONTARIO.

load dropped; fine sand is carried beyond and deposited about the outer margin of the accumulation of boulders and pebbles, and the finer material held in suspension is transported still farther from shore and distributed over the lake bottom. The coarse material is deposited about the mouth of the stream in a semi-circular pile, the base of which is beneath the water and the apex some distance above, where the stream begins to lose velocity. The pile is built out in all directions in which the water has freedom to flow, and a semi-circular or occasionally a truly delta-shaped addition is made to the land.

Fine examples of deltas, built by swift streams adjacent to a precipitous shore, occur on the west side of Seneca lake, New York, near Watkins. In these deltas the action of shore currents from both the north and south is conspicuous, and the deposits have been cut away so as to leave a triangular or markedly delta-shaped outline, but the apex of each "delta" points lakeward, instead of toward the shore as is the normal condition. About the margins of these deltas there are small gravel bars that are frequently looped and enclose lagoons. An active struggle is there in progress between the outflowing streams and the shore currents, which has modified the form of the deltas in the peculiar way just referred to.

A delta advances as fresh material is added to its outer margin, and at the same time the apex of the pile rises and slowly migrates up stream. Such a deposit has a well-defined structure, due to its mode of growth. A radial section made from its apex to any point on its periphery would show three divisions, as is indicated in the following sketch section of a delta built in Lake Bonneville, at Logan, Utah.

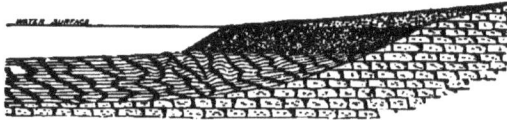

FIG. 7. — SECTION OF A DELTA.

The history to be read in such a section is this: the fine, evenly strati-fied beds beneath the coarse inclined layers are sediments deposited on the lake bottom, but about the margins of deltas they are usually thicker than on neighboring lakeward areas, owing to more rapid depositions from the waters of the delta-forming stream. In some instances a broad, low apron-like deposit of fine sediment is formed about the lakeward margin of the delta proper. As the coarser portion of a delta increases, it advances lakeward and covers the layers of fine sediment previously laid down, and frequently causes them to become folded and wrinkled and occasionally broken and faulted, on account of the weight of material imposed upon them.

The boulders, gravel and sand brought down by a stream are carried to the outer margin of its delta, and roll and slide down its submerged lakeward slope so as to form inclined layers. The angle of inclination of these layers is the angle of stability in water of the material forming them. Where the deposit is mainly of rounded stone and gravel, the angle of slope is in the neighborhood of 30 to 35 degrees, but in some instances is steeper and the structures are unstable and favorable for landslides.

The triangular area shown in the section, above the inclined beds, is the subaërial portion of the delta, built by the stream in meandering

A RECURVED SPIT. GRAND TRAVERSE BAY, LAKE MICHIGAN.

PLATE 12.

over its surface. It is really an alluvial cone, similar to the conical piles of débris so common in desert valleys at the mouths of high grade cañons. It is irregularly stratified, the layers being inclined at a low angle corresponding with the slope of the surface of the structure at the time they were laid down.

The change from the gently sloping and irregularly bedded material of the alluvial portion or cap of the delta, to the steeply inclined and more regularly bedded layers, marks the level of the lake in which the deposit was formed. The outer margin or periphery of the delta, is in a horizontal plane and retains the same position as the delta advances, providing there is practically no change in the level of the lake surface. The surface slope of the cap of the delta, along radial lines from the apex to the periphery, is gently concave to the sky. On recent examples the surface is frequently scored with radiating and branching channels, or "distributaries," left by the changeable stream that built the structure. As a delta increases in size its apex rises and slowly migrates up stream, as already stated, so that in large deltas of high-grade streams the apex is frequently well within the mouth of the cañon through which the drainage is delivered.

In the deltas of low-grade streams, like the Mississippi, the divisions noted above are not readily distinguishable, as the material forming them is fine throughout and the inclination of all the layers is gentle.

Should the surface of a lake be lowered after having stood at a definite horizon for a long period, the terraces, embankments, deltas, etc., formed about its borders become conspicuous features of the exposed land surface and another series of similar forms is at once begun at a lower level. Should another subsidence follow, another series of horizontal lines will be added to the topography of the shores. A rise of a lake causes the submergence of previously formed shore features, and they may become covered with fine sediment or have other wave and current-built structures imposed upon them. Such changes lead to puzzling complications in the records, as has been observed in many instances where lake basins have been emptied and their sides and bottoms laid bare.

Ice-built walls. — In addition to the topographic features characteristic of lake shores thus far noticed, there are others due to the action of ice. In northern latitudes the formation of sea cliffs, terraces, embankments, etc., about the margins of lakes, excepting those of large size, takes place mainly in the summer season. In winter, when most small lakes are frozen over, the expansion of the ice pushes up stones and gravel along shelving shores and forms other topographic features. Another process tending in part in the same direction comes into play in the spring when the ice on a lake becomes broken and is moved by the wind. The action under these conditions is the same that takes place on a much larger scale on the shores of Labrador and other northern lands, where an ice pack is driven on a shelving beach by the force of the wind. Stones and boulders are carried up low lake shores, in the manner here noted, and added to the ridge formed by the winter expansion of the ice. Occurrences of this character have been observed by J. B. Tyrrell on the shore of Lake Winnipegasie.[1] In some instances these ice-built ridges are so marked and appear so much like artificial walls that they are commonly referred to the work of man. In some observed examples in the northern portion of the United States and in Canada, ice-built ridges occur 40 to 50 feet from the water's edge, are 20 feet high and broad enough to furnish convenient roadways.

The formation of ice-built walls about the margins of small northern lakes by ice expansion was first explained by C. A. White.[2] The process has also been clearly stated by Gilbert,[3] in his treatise on the topography of lake shores, from which the following is quoted : —

"The ice on the surface of a lake expands while forming so as to crowd its edge against the shore. A further lowering of temperature produces contraction, and this ordinarily results in the opening of vertical fissures. These admit the water from below and by the freezing of that water are filled, so that when expansion follows a subsequent rise of temperature the ice cannot assume its original position. It consequently increases its total area and exerts a second thrust upon the shore. When the shore is abrupt the ice itself yields, either by crushing at the margin or by the formation of anticlinals (upward folds) elsewhere ; but if the shore is gently shelving, the margin of the ice is forced up the declivity and carries with it any boulders or other loose material about which it may have frozen. A second lowering of temperature does not withdraw the protruded ice margin, but initiates other cracks and leads to a repetition of the shoreward thrust. The process is repeated from time to time during the winter, but ceases with the melting of the ice in the spring. The ice formed the ensuing winter extends only to the water

[1] Geol. and Nat. Hist. Surv. of Canada. Ann. Rep., 1890–91, p. 64 B.
[2] American Naturalist, vol. 2, 1869, pp. 146–149.
[3] Fifth Ann. Rep., U. S. Geol. Surv., p. 109.

SHORE OF LAKE BONNEVILLE, UTAH. (AFTER GILBERT.)

PLATE 13.

margin, and by the winter's oscillation of temperature can be thrust landward only to a certain distance, determined by the size of the lake and
the local climate. There is thus for each locality a definite limit beyond
which the projection of boulders cannot be carried, so that all are deposited along a common line where they constitute a ridge or wall."

Shore walls are not conspicuous about the margin of large lakes for
the reason that they seldom freeze over and also because the winter's ice
work is usually obliterated by the more active waves and currents at
other seasons. They are not formed about deep lakes for the reason that
such water bodies do not become ice-covered, and for the same reason
they do not occur in warm climates.

In this brief sketch of the topography of lake shores, an attempt has
been made to direct attention to the main processes by which the results
have been reached, and to describe briefly the character of some of the
more striking forms produced, without attempting an exhaustive analysis
of the subject. To the reader who would go farther in the studies here
outlined, I most heartily recommend G. K. Gilbert's attractive paper on
the topography of lake shore, in the 5th Annual Report of the U. S. Geological Survey, and the more special volume by the same author on Lake
Bonneville, forming Monograph No. 1 of the publications of the U. S.
Geological Survey.

CHAPTER IV.

RELATION OF LAKES TO CLIMATIC CONDITIONS.

LAKES may be conveniently divided into two great classes, fresh and saline, in reference to the chemical composition of their waters. These two classes have no sharply defined boundary between them, but a complete graduation may be found between the freshest and most saline examples.

A convenient test for determining to which class a lake should be referred is to taste its water. If no saline or alkaline taste is perceptible, it evidently falls in the first class; but if the presence of salts can be determined in this way, it should be referred to the second class.

It is frequently convenient, however, to recognize an intermediate class, or brackish-water lakes, to include water bodies that are slightly saline or alkaline to the taste, but contain only a small fraction of one per cent of mineral matter in solution.

The more pronounced differences in chemical composition, shown by lakes, depend mainly on climatic conditions. Fresh water lakes overflow or else their surplus water escapes by percolation, while saline lakes are without outlets. Exceptions to this rule may occur, but they are accompanied by unstable conditions, and the presence of an outlet to a saline lake or its absence in the case of a fresh lake, are temporary phases that have not continued long enough to bring about the changes toward which they tend.

Fresh lakes occur principally in humid regions, while saline lakes, with the exception of those formed by the isolation of bodies of sea water, are confined to regions of small rainfall. Whether a lake shall overflow or not, depends ordinarily on the relation of the rainfall over its hydrographic basin to evaporation from the lake surface. As lakes frequently receive the water of fissure springs, the sources of which may be far distant, it will be more exact to say that whether a lake held in an impervious basin shall overflow or not, depends on the ratio of the amount of water contributed to it to the amount evaporated from its surface. If the inflow is in excess of evaporation, the water will rise and its area increase until an equilibrium is established or until an outlet is found.

When evaporation counterbalances the inflow for a long period, the waters are concentrated and become charged with mineral matter, for the reason that all streams and springs contain foreign substances in solution which are left when evaporation takes place.

It has been found by observation that in regions where the topographic conditions are favorable, a rainfall of about 20 inches per year, and an evaporation from lake surfaces in excess of 50 inches per year, is frequently accompanied by the formation of lakes that do not rise sufficiently to find an outlet. When the difference in the direction indicated between precipitation and evaporation is still greater, or when the area from which a lake receives the drainage is small in reference to the area where a lake would naturally form, desiccation may be complete and permanent lakes rendered impossible.

Whether a lake shall be fresh or saline depends, therefore, on climatic conditions and on the configuration of its hydrographic basin.

FRESH LAKES.

Material in Solution. — As all fresh lakes may be considered as the expansions of streams, their chemical composition is indicated where the actual lake waters have not been analyzed, by the composition of the streams flowing to or from them. It follows, therefore, that the average composition of the waters of fresh lakes would be shown with considerable accuracy, by the average composition of the principal rivers in the region where they occur.

Analyses of the waters of 20 of the principal rivers of the United States have shown that they contain on an average 0.15044 part per thousand of total solids in solution, of which 0.056416 part per thousand is calcium carbonate. This may be taken as the average composition of the fresh lakes of this country, but more particularly of those in the humid regions.

In a table of 48 analyses of European river waters given in Bischof's Chemical Geology, the average of total solids in solution is 0.2127 and the average of calcium carbonate 0.1139 part per thousand. From the analyses of the waters of 36 European rivers given in Roth's Chemical Geology, including some of those mentioned by Bischof, the average of total solids is 0.2033 and of calcium carbonate 0.09598 part per thousand.

In both American and European rivers, as determined from the above data, the average of total solids in solution is 0.1888 and of calcium carbonate 0.088765 part per thousand. These figures may be safely assumed to represent the average amount of impurities carried by normal streams, and consequently indicate the character of the lakes to or from which they flow. The drainage in mountainous regions, especially where supplied by melting snow and ice, may be purer than these figures indicate ; while in arid regions, where efflorescent salts frequently whiten the surface, the streams are more highly charged with saline matter than when the rainfall is abundant. It is to be observed that material carried by streams in suspension is not included in the above considerations.

The reader may, perhaps, conclude from the figures just given that the percentage of saline matter carried in solution by ordinary streams is unimportant and of but little significance in connection with the study of lakes. It is true that the amount of foreign matter in solution in a few gallons of river water is small, but where the volume of rivers is considered the amount of solid substances carried by them in solution, even in a single year, becomes truly startling. Knowing the volume of a stream and the percentage of mineral matter it contains, one can readily compute the weight of the matter it carries in solution in a definite time. This computation has been made for a few American rivers.[1]

The average flow of Croton river, New York, is 400,000,000 gallons daily. In this volume of water there are 183 tons of mineral matter in solution, of which 47 tons are calcium carbonate.

The Hudson carries daily about 4,000 tons of matter in solution, of which more than 1,200 tons are calcium carbonate.

The Mississippi carries to the Gulf of Mexico in a single year about 113 million tons of mineral matter in solution, of which over 50 million tons are calcium carbonate.

These estimates are only approximately correct as they depend in most instances on a single analysis and on a small number of measurements of volume.

The invisible loads carried by rivers are not only of interest in connection with the study of lakes, more especially of saline lakes, but open a wide field of research in reference to the chemical denudation of the land, the composition of ocean waters, and the source of the material, more particularly of the calcium carbonate, secreted by marine plants and animals. Into this broader domain, however, to which our subject leads, we may not now enter.

---

[1] The data from which the facts here stated were obtained, as well as similar information concerning other streams, is given in Monograph No. 11, U. S. Geol. Surv., pp. 172–175.

TYPES OF FRESH LAKES.

Of the tens of thousands of fresh lakes scattered over North America, and especially abundant in the previously glaciated, northeastern portion of the continent, or forming a part of the more impressive scenery of the Cordilleran region, many might be selected as types. Attention will be confined, however, to the Great Lakes, drained by the St. Lawrence, Lake Tahoe, California, and Lake Chelan in the State of Washington.

The Laurentian lakes. — The group of great lakes drained by the St. Lawrence, as is well known, contain the most magnificent examples of fresh water-bodies now existing on the earth. Lake Superior still retains its position as the largest sheet of fresh water known, although the more recent discovery of Lake Victoria Nyanza has brought a rival into the field. This African lake is estimated to have an area of about 18,000 square miles, which is 12,000 square miles less than the area of the great American lake; but when an actual survey shall have been made, it is possible that this difference will be materially decreased.

While Lake Superior exceeds all other fresh lakes in extent, it ranks second among terrestrial water-bodies, for the reason that the Caspian Sea is the largest sheet of water not in open communication with the ocean, now existing. The Caspian is saline, however, and falls in the second great division of lakes here recognized.

The origin of the basins of the Laurentian lakes has been referred to in Chapter I. in connection with the action of glacial agencies in obstructing drainage; an account of their past history is given in advance in discussing the Pleistocene lakes of the same region; at present attention will be confined to some of the more interesting features of the existing lakes.

The U. S. Lake Survey. — A survey of the Laurentian lakes was made by the Corps of Engineers, U. S. Army, between 1841 and 1881, and is known as the U. S. Lake Survey.[1] On the maps or chart published by that survey, the outlines of the shores of the lakes and of their connecting waters are given, together with the topography of a narrow strip of the adjacent land; the depth of water, character of bottom, etc., as

[1] Report upon the Primary Triangulation of the U. S. Lake Survey, by Lieut.-Col. C. B. Comstock, Washington, 1882.

determined from thousands of soundings, is also indicated. This excellent survey is the basis of nearly all accurate information now accessible concerning the physical features of the lakes in question, and has been freely used in compiling the following statements.

Owing to changes in the rivers connecting the various Laurentian lakes and in bays and navigable channels, and also on account of the many harbor and canal improvements that have been made, a new survey of portions of these lakes has been found necessary, and is now in progress under the direction of Gen. O. M. Poe.

The area of the Laurentian lakes has been determined with approximate accuracy from measurements made on the maps of the U. S. Lake Survey. The results of these measurements by different individuals vary somewhat, but those published by L. Y. Schermerhorn[1] are here adopted.

AREA OF THE LAURENTIAN LAKES IN SQUARE MILES.

	WATER SURFACE.	WATER SHED.	HYDROGRAPHIC BASIN.
Lake Superior	31,200	51,600	82,800
St. Mary's river	150	800	950
Lake Michigan	22,450	37,700	60,150
Lake Huron and Georgian Bay . . .	23,800	31,700	55,500
St. Clair river	25	3,800	3,825
Lake St. Clair	410	3,400	3,810
Detroit river	25	1,200	1,225
Lake Erie	9,960	22,700	32,660
Niagara river	15	300	315
Lake Ontario	7,240	21,600	28,840
Total	95,275	174,800	270,075

The volume of water flowing through the rivers draining the various lakes is on an average as follows:

	CUBIC FEET PER SECOND.
St. Mary's river, the outlet of Lake Superior . .	86,000
St. Clair river, the outlet of Lakes Huron and Michigan	235,000
Niagara river, the outlet of Lake Erie . . .	265,000
St. Lawrence river, the outlet of Lake Ontario . .	300,000

[1] "Physical Features of the Northern and Northwestern Lakes," Amer. Jour. Sci., 3d sec., vol. 33, 1887, pp. 278-284.

The mean elevation of the surfaces of the Laurentian lakes above the sea, their maximum depth, etc., as shown by soundings, are as follows :

MEAN ELEVATION AND MAXIMUM DEPTH, ETC., OF THE LAURENTIAN LAKES.

	MEAN ELEVA-TION.	APPROXIMATE MEAN DEPTH.	MAXIMUM DEPTH.	DEPTH OF BASIN BELOW SEA LEVEL.
Lake Erie	573	70	210	
Lake Huron	581	250	730	149
Lake Michigan . . .	581	325	870	289
Lake Ontario	247	300	738	491
Lake Superior	602	475	1,008	406

The average discharge of the lakes is stated by Schermerhorn to be double that of the Ohio and nearly equal to one half the discharge of the Mississippi. The area of the Laurentian basin is a third larger than the hydrographic basin of the Ohio, or about a fifth of the combined areas of the basins of the Mississippi and its affluents. The outflow of the St. Lawrence basin is slightly less than half its rainfall, while on the Mississippi and Ohio the discharge is about a fourth of the rainfall. If the average discharge of the Laurentian lakes passed through a river one mile wide with a mean velocity of one mile per hour, such a river would have a depth of 40 feet from shore to shore.

The volume of water in the Laurentian lakes is about 6,000 cubic miles, of which Lake Superior contains somewhat less than one half. Perhaps a better idea of this volume may be obtained when it is said that it is sufficient to sustain Niagara falls in their present condition for about *100 years.*

The mean annual rainfall of the St. Lawrence basin is about 31 inches ; and the mean depth of water evaporated from the surfaces of the lakes, between 20 and 30 inches.[1] The amount of precipitation on the water surface is, therefore, nearly compensated by the amount evaporated from the same area.

Chemistry of the waters of the St. Lawrence. — The composition of the waters of the Laurentian lakes is shown with approximate accuracy

[1] Thomas Russell, "Depth of Evaporation in the United States," Monthly Weather Report, U. S. Signal Office, Sept. 1888.

by an analysis of the water of St. Lawrence river taken near Montreal.
This analysis may also be considered as representing very nearly the composition of the material carried in solution by the lakes and rivers of the
more humid portions of North America.[1]

ANALYSIS OF THE WATER OF ST. LAWRENCE RIVER.

BY T. STERRY HUNT.[2]

INGREDIENTS.	PARTS IN A THOUSAND.
Sodium, Na	.00513
Potassium, K	.00115
Calcium, Ca	.03233
Magnesium, Mg	.00585
Chlorine, Cl.	.00242
Carbonic acid, CO_3	.06836
Sulphuric acid, SO_4	.00831
Phosphoric acid, HPO_5	trace
Silica, SiO_2	.03700
Alumina, Al_2O_3	trace
Oxide of iron, FeO	"
Oxide of Manganese, MnO	"
Total	0.16055

Taking the volume of the St. Lawrence at 300,000 cubic feet per
second, the computed discharge of Lake Ontario, it follows from the above
analysis that approximately 1.5 tons of mineral matter in solution is transported by it per second, or about 50 million tons annually.

Erosion of the lake shores. — The shores of the Laurentian lakes
are being eroded at many localities, and the material thus removed deposited, in part, on other portions of the coast so as to add to the land
area. Some information in this connection has been compiled by Charles
Crosman,[3] but much additional data is required before general conclusions
of value can be reached.

The average annual recession of the sea-cliff along the west side of
Lake Michigan, as determined by Prof. Edward Andrews from a somewhat extended series of observations, is stated to be about 5 feet; with a

[1] Analyses of the water of 20 rivers of the United States and Canada may be found in
Monograph No. XI, U. S. Geological Survey, Table A.

[2] Geological Survey of Canada, 1863, p. 567.

[3] "Chart of the Great Lakes." Published at Milwaukee, Wisconsin.

maximum at certain localities, of 16 feet. In the neighborhood of Cleveland, Ohio, the mean recession of a line of prominent sea-cliffs in boulder clay, for a period of 40 years, has been about 6 feet per annum.

Observations at less favorable localities show a similar retreat of other portions of the lake shores, but definite quantitative observations have seldom been recorded. Enough is known in a qualitative way, however, to show that important changes in the outlines of these lakes are in progress. The waste of the shore, resulting in a broadening of the surfaces of the lakes, is compensated in part by the deposition of the material removed on adjacent area so as to extend the land lakeward, as, for example, at the south end of Lake Michigan, where beaches and large sand dunes have been formed, and are still encroaching on the lake. Observations made by the writer at various localities about the shores of the lakes, together with the reports of others, show conclusively that the process of broadening the lakes by the erosion of their shores is progressing more rapidly than areas are being reclaimed by deposition, and therefore that they are becoming shallower.

Commerce and fisheries. — The importance of the Laurentian lakes as highways of commerce is too well known and is too extended a subject to receive treatment at this time, even if it fell within the scope of the present discussion. Some idea of the magnitude of the commerce on these inland waters may be had, however, from the reports of the operation of the Government locks at Sault St. Marie, which show that 11,557 vessels passed through them during the year ending June 30, 1892, carrying over 10 million tons of freight. The great importance of the commerce of the Laurentian lakes will be better appreciated, by those who are not familiar with it, when it is compared with the traffic of the Suez Canal. In 1889, the latest date at which comparative data are at hand, nearly three times as many vessels passed through the locks at Sault St. Marie as through the Suez Canal, although the latter is open for navigation throughout the entire year. The tonnage during the same year was 7,221,935 at the "Sou," as against 6,783,189 for the Suez Canal. The importance of the carrying trade of the Great Lakes is also shown by the fact that the tonnage of vessels constructed on them each year for several years, has been about equal to that of all the vessels built on the Atlantic, Pacific, and Gulf coasts. Still more striking is the fact that the amount of goods carried each year on these inland waters, is far in excess of the entire clearances of all the seaports of the United States, and several million tons in excess of the combined foreign and coastal trade of London and Liverpool.

The demand for still better facilities for inter-lake communication has led to the construction of still larger canals and locks, and now improvements are nearly completed which will allow vessels drawing 21 feet of water to pass from Buffalo to Duluth. It is expected that when this improvement is made the trade between Lake Superior and the more southern lakes will be doubled in a few years. Far-reaching plans for connecting this important commercial industry with ocean highways are under consideration, and must find consummation in the near future.

The fisheries of the Laurentian lakes is another subject of great practical importance, as they are the most extensive lake fisheries in the world. The lakes abound in trout, whitefish, and other food fishes, and their shores are dotted with fishing villages, in which a hardy population, skilled in all that pertains to their calling, are living their humble but useful lives, and gaining an experience which well fits them for naval service should their aid be called for. The importance of these inland fisheries has received tardy recognition in comparison with the similar industries of the sea border, but they are a substantial element of national wealth and claim the most careful attention and guidance of both state and national legislators. The reports of the U. S. Fish Commission show that over ten thousand persons are engaged in this industry ; that the capital invested is in excess of four and one-half millions of dollars ; and that a hundred million pounds of fish are secured each year, which bring to those actually engaged in the work more than two and one-half millions of dollars.

It may be noted as an item of interest in connection with the physical history of the Laurentian basin, that in lakes Superior and Michigan crustaceans and fishes have been found that are believed to be identical with living marine forms. These are thought by some persons to indicate that the lakes in which they occur were formerly in open communication with the ocean. Considerable evidence, derived from a study of the former extent of the lakes, and of the fossils in the sediments of previous water-bodies in the same basins, do not seem to confirm this conclusion, however, and further study of the habits and means of migration of the species referred to, is necessary before their presence in inland waters can be satisfactorily accounted for.

The movements of the waters of the Laurentian lakes and a few facts respecting their temperature and their influence on the climate of the adjacent land have already been referred to in preceding chapters. Scarcely more than a beginning of their physical study has been made, however, and it is to be hoped that they may soon receive the attention in this direction they so well deserve.

Mountain lakes. — No account of the lakes of North America is complete that does not include some notice of the thousands of basins amid the northern Appalachians, and in the Cordilleras, in which the most magnificent scenery of this continent is reflected. These lakes are of all sizes, from mere tarns across which one might spring with the aid of an alpenstock, to broad plains of blue, many square miles in area, and worthy of comparison with the most beautiful mountain lakes of other lands. Of this attractive class of lakes special attention can only be given at present to two examples which are destined to be widely known on account of their many charms. I refer to Lake Tahoe, embosomed among the peaks of the Sierra Nevada, and lying partially in California and partially in Nevada; and to a lake of a different character but not less magnificent, situated in the Cascade mountains, in the State of Washington, and known as Lake Chelan.

Lake Tahoe. — This "gem of the Sierra" is situated at an elevation of 6200 feet above the sea and is enclosed in all directions by rugged, forest-covered mountain slopes which rise from two to over four thousand feet above its surface. Its expanse is unbroken by islands and has an area of between 192 to 195 square miles. Its diameter from north to south is 21.6 miles and from east to west 12 miles.

On looking down on Lake Tahoe from the surrounding pine-covered heights, one beholds a vast plain of the most wonderful blue that can be imagined. Near shore, where the bottom is of white sand, the waters have an emerald tint, but are so clear that objects far beneath the surface may be readily distinguished. Farther lakeward, the tints change by insensible gradation until the water is a deep blue, unrivaled even by the color of the ocean in its deepest and most remote parts. On calm summer days, the sky with its drifting cloud banks and the rugged mountains with their bare and usually snow-covered summits, are mirrored in the placid waters with such wonderful distinctness and such accuracy of detail, that one is at a loss to tell where the real ends and the duplicate begins. While floating on the lake in a boat, the transparency of the water gives the sensation that one is suspended in mid air, as every detail on the bottom, fathoms below, is clearly discernible.

In experimenting on the transparency of the waters, Professor John LeConte found that a white disc 9.5 inches in diameter, when fastened to a line and lowered beneath the surface, was clearly visible at a depth of 108 feet. It is to be remembered that the light reaching the eye in such an experiment traverses through water twice the distance to which the disc is submerged, or in the experiment referred to, 216 feet. The only instance in this country in which waters have been found to be more transparent is in the great limestone-water springs of Florida.

Soundings made in Lake Tahoe by LeConte, as already stated, gave a maximum depth of 1645 feet, but a more detailed survey may possibly discover still more profound depths. Those measurements show that the lake, with the exception of Crater lake, Oregon, is the deepest inland water-body in America yet sounded, and exceeds the depth of any of the lakes of Switzerland, but is not so deep as lakes Maggiore and Como on the south side of the Alps.

The temperature observations made in Lake Tahoe previously referred to, furnish an illustration of the fact that deep lakes, even when situated at a high elevation and subject to low winter temperatures, do not freeze. The surface waters are cooled in winter and descend, while warmer waters from below rise and take their place, thus establishing a circulation, but the body of water is so great that its entire mass never becomes cooled sufficiently during the comparatively short winters to check the upward circulation and allow ice to form. At the greatest depth reached the temperature was 39.2° F., which is the temperature of fresh water at its maximum density; and from more extended observation in other lakes, the water is believed to retain this temperature throughout the year.

Lake Tahoe is situated at such an altitude that its shores are bleak and inhospitable during a number of months each year. For this reason it is probable that it will never be selected as a place of continued residence by any considerable number of families, but during the summer, when the adjacent valleys are parched by desert heat, the air in the lake-filled valley is cool and bracing ; it then furnishes a charming retreat for the dwellers of the cities of the Pacific coast, as well as for more distant wanderers. As a place for summer rest and recreation it is second to none of the popular resorts of the United States or Canada.

The waters of Lake Tahoe overflow through the Truckee cañon and form a bright, swift-flowing stream, which finds its way to Pyramid and Winnemucca lakes, situated 2400 feet lower, in the desert valleys to the north. The waters when starting on their troubled journey are as pure and limpid as the melting snows of mountain valleys can furnish. Analyses show that they contain only 0.0750 part per thousand of mineral matter in solution, but the lakes into which they flow and of which they form almost the sole supply, are alkaline and saline owing to long concentration.[1]

An example of an isolated drainage system is here furnished, embracing the cool summits of lofty mountains where the moisture of the atmosphere is condensed ; a mountain reservoir where the waters are stored ; a swift, clear stream formed by the overflow of the reservoir ; and the bitter lakes where the stream empties and from which there is no escape except by evaporation. Such an attractive field for geographical study should not be long neglected. A careful investigation of the various problems here assembled in narrow bounds, would form a thesis of unusual interest. Will not some student or some class of students in our universities tell the world what the mountains and streams in this fascinating region are doing, explain how the present conditions came into existence, and point out the results towards which they are tending?

Lake Chelan. — Our second example of mountain lakes, selected from the large number that shimmer in the sunlight amid the highlands of the Far West, lies hidden in the embrace of the eastward-reaching spurs of the Cascade mountains in the State of Washington, and until recently was so remote from the paths ordinarily followed by man, that its very name will sound strange to many of my readers.

Where Columbia river crosses the arid region between the Rocky mountains and the Cascade range, making a vast sweep about the northern and western margins of an ancient lava flood, it washes the bases of the mountains to the west and receives the tribute of a number of lakes, fed by the melting snow on the higher portions of the range. One of these lakes, named in honor of Chelan, an Indian chief of considerable local renown, whose village stands on its shore, empties into the Columbia through a deep tortuous gorge of recent origin and sends a swift stream of clear, greenish-tinted water about two miles long, to join the great river in the adjacent cañon. The lake is a narrow, river-like sheet of water, with gentle windings, extending westward from the Columbia, seventy miles into the mountains, and is bordered on either hand by a continuous series of rugged peaks that rise from five to over seven thousand feet above its surface. The deep, narrow, trench-like valley, now partially

---

[1] For analyses of the waters of these lakes, see p. 72.

water-filled, continues beyond the head of the lake for a distance of at least twenty-five miles, becoming more and more wild and rugged as it nears the heart of the highlands. The total length of this remarkable valley is not less than one hundred miles, and its width at the level of the lake seldom exceeds four miles.

The sounding line has shown that Lake Chelan is over eleven hundred feet deep, but its full depth remains to be determined. In several soundings made by the writer in its central and western portions, no bottom was reached at the depth indicated. The surface of the lake is but 950 feet above the sea, so that the bottom of the trough is below sea level.

Where the clear water of the lake washes the precipitous walls enclosing it there is no beach, and scarcely a trace on the rocks to show that it has altered the topography of the shores. The present conditions were initiated at such a recent date that, practically, the only changes they have produced are at the eastern end of the lake, where it emerges from the rocky defile of the mountains and for a short space expands between comparatively low shores of gravel and sand. In this region high terraces mark the former level of the water surface.

How the great gash in the mountain, fully one hundred miles long, and now filled for more than a thousand feet in depth by the lake, was formed, is not easy to explain. Previous to the birth of the present lake the valley was occupied by a large glacier which flowed through it and joined another great ice stream in the cañon of the Columbia. The ice smoothed the precipices of rock and piled up moraines on the more gentle slopes at the east end of the valley, but that the main depression existed before the glacial invasion is evident and is in harmony with the histories of many other valleys in the Cordilleran region. The valley has a still more ancient history, and in Tertiary, or in part perhaps in pre-Tertiary times, was excavated in the hard granite, now seen in its enclosing walls, by the slow wear of streams. It is a stream-cut channel, but where the stream rose that did the work, or whence it flowed, remains to be determined by a careful study of all the facts bearing on the problem.

It has been the writer's fortune to pitch his camp on the borders of both Lake Tahoe and Lake Chelan. As the scenery of each is conjured up in revery, it is difficult to decide which is the more remarkable or which should have the first rank among the mountain lakes of America. Each lake is surrounded by forest-covered mountains of majestic proportions and rich and varied details ; the waters of each lake are clear and deep in color, or varied by silvery reflections and iridescent tints where the not too gentle mountain winds touch their surfaces; in each instance the scene is fresh and unmarred, and has the charm of remoteness so welcome to many who are weary with the ways of men.

At Tahoe the views are wide and far-reaching. The shaggy mountains are picturesquely grouped about the central plain of waters and the scene is open and, for a mountain stronghold, mild and pleasing.

At Lake Chelan the scenery is wild and rugged. The narrow stream-like sheet of water, with gently curving shores, extends far into the mountains and cannot be comprehended at a glance. Each view, as one ascends the lake, gives suggestions of something still more grand beyond. Each turn reveals hidden beauties that entice one on and on. The bordering mountains become more and more rugged, as we venture farther into their embrace. Each newly discovered peak is higher and more imposing than its predecessor; until at the head of the lake, the most lofty summits of the range, usually white with snow, can be seen far up the gorge beyond where boats can go. The narrow valley bottom beyond the lake is filled with majestic trees and a rich profusion of lower vegetation of almost tropical density; the dark vine-entangled forest seems striving to conceal some mysterious shrine farther within the heart of the mountains. A clear, swift stream flows silently beneath the deep shade of the broad-leaved sycamores; and from far within the hidden recesses of the valley, the echoes of unseen cataracts come faintly to the ear. What wonders exist in the upper portion of the valley are not known, as they have been seen by only a few white men and have never been described.

All of the surroundings of this wonderful lake are so fresh and speak so strongly of the untamed beauties of Nature in her wildest moods, that a visit to the region has the zest and fascination of entering an undiscovered country, where each step takes one farther and farther into the unknown.

The vegetation of the Cascade mountains is far more luxuriant and varied than the flora of the Sierra Nevada. In every nook and corner one is surprised and charmed with the rank luxuriance of the gracefully bending ferns, or the profusion and brilliancy of the flowers. On the higher slopes, between the forests and the bare summits of the cloud-capped peaks, the angles of the rock are softened by luxuriant mosses and lichens, and the gray of the cold granite is brightened by Alpine blossoms.

Tent life on the shore of either Lake Tahoe or Lake Chelan is delightful. Each lake has its own peculiar charms, but their influences on the mind are different. One or the other will be declared the more attractive according to the temperament of the person who yields himself to their influences. Each is poetic, and will weave a web of golden fancies in the mind of its admirer, which will be as nectar to his thoughts when his feet tread other and less inspiring paths.

Owing to the very moderate elevation of Lake Chelan, its climate is mild throughout almost the entire year, and is delightful from early spring to late autumn. Since the building of the Great Northern railroad, this charming lake of the Cascades is quite accessible. The traveler leaving the railroad at Wenatchee, may ascend the Columbia by steamer, to Chelan Crossing, a distance of about forty miles, and thus see something of the great river of the Northwest. From Chelan Crossing, a ride, or preferably a walk of two miles, will bring the visitor to Chelan "City" as a unique group of several hundred "claim shanties" is termed. The houses in this silent city were built simply for the purpose of acquiring some sort of a title to the land on which they stand and were never intended for habitation. The generous hospitality of the sparse population in this frontier town makes up for their lack of numbers. Every visitor who comes to see the beauties of the lake and mountains, of which the dwellers of the region are justly proud, will be welcomed.

On the lake there are small steamers, which make regular trips to its head, and boats for sailing and fishing. The trout in the lake are abundant and unusually fine. Mountain goats inhabit the higher mountains, and afford sport equal to the chamois chase. Small hotels have been built on the shores of the lake for the accommodation of summer tourists, fishermen, and hunters. I mention these details for the purpose of assuring the reader that he will find traveling easy and agreeable, if he wishes to verify what has been stated in reference to the attractions of one of the wildest and grandest lakes in America.[1]

Only two examples of the mountain lakes of America have been referred to, for the reason that the space at command does not permit even the mention of the hundreds of charming examples, many of them of greater size and in their milder fashion as attractive as those of the Sierra Nevada and Cascade mountains, which add variety and beauty to the New England States, New York, etc. Extending our survey to Canada, a still

[1] A more complete account of the region about Lake Chelan than can be given at this time, may be found in a report on the Upper Columbia River by Lieut. T. W. Symons; 47th Congress, 1st session, Senate Executive Doc. No. 186, Washington, 1882; and in a report by the author, on a Geological Reconnoissance in Central Washington, U. S. Geol. Surv., Bulletin No. 108.

greater host of inland water bodies of almost every variety imaginable, attract the attention and cause our pen to linger; but here again we can only say that they belong to a great class of which types have been briefly described.

SALINE LAKES.

Saline lakes are formed principally in two ways. First, by the isolation of bodies of sea water, as where a rise of the land cuts off an arm of the ocean, or sand bars or coral reefs enclose lagoons. Second, by the concentration by evaporation of ordinary river waters in enclosed basins. The first are of oceanic and the second of terrestrial origin.

Saline lakes of oceanic origin. — There are no conspicuous examples of this class of lakes in North America, although lagoons cut off from the ocean by sand bars do occur, especially along the southern Atlantic coast.

A large lake of salt water that was isolated from the ocean by a rise of the intervening land formerly occupied the valley of Lake Champlain, but has been freshened and its surface lowered by overflow.

The type of saline lakes which were formerly arms of the ocean is furnished by the Caspian sea, the largest body of inland water known. The observations of many travelers have shown that this sea has been divided from the ocean by the elevation of the intervening land. The climate of southwestern Asia is arid, and over large areas evaporation is in excess of precipitation. For this reason the Caspian has contracted its borders, in spite of the large contribution of water delivered to it by the Volga and other streams.

There is evidence in the chemical composition of the waters of the Caspian, and in the topography of land separating it from the Black sea, to indicate that at first it was freshened by overflow, as in the case of the ancient lake of Champlain valley, and that its present salinity has resulted principally from the concentration of river waters. It may be considered, therefore, of oceanic or of terrestrial origin as one chooses.

The Caspian is 180,000 square miles in area, or nearly six times the size of Lake Superior. Its maximum depth is in the neighborhood of 3,000 feet, and exceeds the depth of any other lake known. It is without outlet. Its waters contain 6.294 parts in a thousand of mineral matter in solution, consisting principally of sodium chloride and magnesium sulphate. The waters of the ocean, it will be remembered, contain, on an average, 34.4 parts per thousand, or, in round numbers, 3.5 per cent.

One of the most instructive features connected with the Caspian is the manner in which it loses its saline constituents by discharging into a secondary basin, where the waters are still more highly concentrated. On its eastern shore there is a deep bay or gulf known as Karabogaza, which is nearly shut off from the main water-body by intervening sand bars, and receives its only influx through an opening in the bar about 140 yards broad and 5 feet deep. The water escapes from Karabogaza solely by evaporation, and is replaced by a current from the Caspian which has been estimated by Von Baer to carry 350,000 tons of saline matter daily from the sea to the gulf. The waters of the gulf have reached the point of saturation for common salt, and precipitation is taking place. These peculiar conditions are of great interest, not only in showing how deposits of salt may accumulate, but also in illustrating the manner in which an enclosed lake may deposit a large part of its foreign matter without the entire water-body becoming highly concentrated.

Saline lakes of terrestrial origin. — The existence of lakes of this class depends upon a combination of topographic and climatic conditions. The basins they occupy may originate in almost any of the various ways enumerated in Chapter I. As a rule the lakes of this class in North America occupy depressions formed by movements in the earth's crust which have cut off large areas from free drainage to the sea. Such enclosed basins, however, can only continue in regions where the rainfall is small, for the reason that if precipitation were in excess of evaporation, they would become filled to overflowing. The most favorable conditions for the formation of inland saline lakes are found where high mountains discharge their drainage into basins where the climate is arid. A region of condensation of atmospheric vapors and a region of concentration by evaporation are thus supplied, which supplement each other.

The saline lakes of arid regions are peculiarly sensitive to climatic changes, and undergo many fluctuations. When the mean annual influx and the mean annual loss by evaporation are nearly evenly balanced, lakes frequently exist only during the rainy season, and disappear entirely during the hotter portions of the year, leaving broad, smooth mud plains. Plains of this character are a characteristic feature of the arid region of North America, and are known in Mexico and in the southwestern part

SALINE AND ALKALINE LAKES IN THE ARID REGION.

of the United States as *playas*. It is convenient to adopt this name, and call the temporary water-bodies to which playas owe their origin, *playa lakes*. These lakes may be formed by a single shower and disappear in a few hours, or they may endure for a series of years and only be evaporated to dryness during seasons of exceptionally low rainfall or unusually active evaporation.

When enclosed lakes of arid regions are more permanent, they fluctuate in volume, and consequently in extent and in density, from season to season, and are so sensitive to climatic changes that they show marked variations when ordinary weather observations, taken at a limited number of localities in their neighborhood, fail to indicate analogous changes in atmospheric conditions.

The terrestrial saline lakes of North America are confined to the arid region of Mexico and the United States, although small pools of alkaline water do occur on the great plains in the sub-humid region east of the Rocky mountains both in the United States and Canada. The saline lakes of the United States are confined almost entirely to Utah and Nevada and adjacent portions of the Great Basin. The distribution of some of the more important lakes here referred to, is indicated on the accompanying map forming Plate 14. The chemical composition of their waters is shown in the table on page 72.

Chemical precipitates. — The deposition of mechanical sediments, as clay and sand, in lake basins has already been referred to. This takes place in all lakes without special reference to their chemical composition. When lake waters become concentrated by evaporation, however, the material contributed to them in solution may be precipitated, and either mingle with the mechanical sediments or form deposits of purely chemical origin. Chemical precipitates, like mechanical sediments, may furnish evidence of important changes in a lake's history, and are also frequently of great interest on account of their commercial value.

As already seen, enclosed lakes are constantly receiving contributions from streams, springs, and rain, but do not overflow, the influx being counterbalanced by evaporation. This assures us that in the earlier stages of their history, at least, the amount of saline matter held in solution must increase from year to year and from century to century. This process continuing, a time is eventually reached when the waters will be saturated with one or more of its saline constituents and precipitation begin. Waters

ANALYSES OF THE WATERS OF SALINE AND ALKALINE LAKES IN THE ARID REGION.[1]

[Expressed in parts in 1,000.]

CONSTITUENTS.	ALBERT LAKE, OREGON.[2]	GREAT SALT LAKE[3] UTAH.	HUMBOLDT LAKE, NEVADA.	SODA LAKE, NEVADA.	MONO LAKE,[2] CALIFORNIA.	OWENS LAKE, CALIFORNIA.[3]	PYRAMID LAKE, NEVADA.	SEVIER LAKE, UTAH.	SOAP LAKE, WASHINGTON.	WALKER LAKE, NEVADA.	WINNEMUCCA LAKE, NEVADA.
Sodium, Na	11.215	49.690	.27812	40.919	18.837	26.836	1.1796	28.840	10.504	.85535	1.2970
Potassium, K	.521	2.407	.06083	2.357	.920	1.518	.0733	...	...	Trace.	.0086
Calcium, Ca	...	.255	.01257	...	.109	.013	.0089	.118	Trace.	.02215	.0196
Magnesium, Mg	...	3.780	.01618	.245	.054	.001	.0797	2.600	.011	.03830	.0173
Lithium, Li	...	Trace.	Trace.	...	...	...	...	...	...	...	...
Chlorine, Cl	13.055	83.946	.29545	40.851	11.582	18.214	1.4300	45.500	3.526	.58375	1.6934
Bromine, Br	...	Trace.	...	...	...	...	...	...	...	...	...
Carbonic acid, CO_3	9.199	...	.20126	16.858	13.100	18.265	.4990	...	9.625	.47445	.3458
Sulphuric acid, SO_4	.685	9.858	.03010	11.857	6.384	7.067	.1822	9.345	4.362	.52000	.1333
Phosphoric acid, H^3PO_4	...	...	.00069	...	...	...	...	...	...	...	...
Nitric acid, NO_3	...	...	...	.286	.153	.346	...	...	...	...	...
Boracic acid, B_4O_7	.224	Trace.	Trace.	.278	.067	.207	.0334	...	.113	.00750	.0275
Silica, SiO_2	...	...	.03250	...	.003 }	.023 }	...	...	...	...	...
Alumina, Al_2O_3	...	...	...	...	{	.013	...	...	...	...	...
Iron, sesquioxide, Fe_2O_3	...	...	...	...	...	...	...	...	Much.	...	...
Organic matter	...	...	...	...	...	...	...	...	.053	...	...
Hydrogen, in bicarbonates	.056	...	...	...	.049	.059	...	...	...	...	...
	37.985	149.936	.92860	113.647	51.168	72.595	3.4861	86.403	28.194	2.50150	3.6025

[1] Compiled principally from Table C, U. S. Geol. Surv., Monograph No. XI.
[2] Analyses by T. M. Chatard, Am. Jour. Sci., ser. 3, vol. 37, 1888, pp. 146-150.
[3] In 1869.

holding a number of salts in solution, when slowly evaporated, do not deposit them in a homogeneous mass, but in successive layers of varying composition. As the order in which different salts are deposited varies with the composition of the waters, it is safe to say that in no two lakes is the succession of saline deposits formed on evaporation apt to be the same. Disregarding for the present the reaction of the various salts upon each other, it is evident that in the evaporation of natural waters the order in which the contained salts will be precipitated is inversely as the order of their solubility. For example, a salt which requires a large amount of water for its solution, or, in other words, is sparingly soluble, will reach its point of saturation and commence to crystallize out as evaporation progresses, previous to the deposition of a more soluble salt. To illustrate : it has been found that calcium carbonate requires about 10,000 times its weight of water, saturated with carbon dioxide, for its solution; while calcium chloride is deliquescent, and dissolves in about its own weight of water. In an enclosed lake to which streams and springs are bringing these two salts in equal quantities, and in which evaporation equals or exceeds the supply of fresh water, it is evident that the calcium carbonate would reach its point of saturation and commence to separate long before the waters had become · rich in calcium chloride. In fact, owing to the deliquescent nature of the chloride, natural evaporation seldom proceeds far enough to cause its precipitation. The early deposition of calcium carbonate, when natural waters are concentrated by evaporation, is rendered the more certain for the reason that it is by far the most abundant salt found in surface waters.

The fact that various salts are deposited in a regular succession when mineral waters are evaporated, is of great service in separating certain ones in a pure state by the method known as fractional crystallization. In evaporating the brines of Syracuse, New York, the precipitation of ferric oxide and of calcium sulphate, or gypsum, is first secured by moderate concentration ; the brine is then conducted to lower vats and evaporation continued until the sodium chloride, or common salt, has mostly crystallized and fallen to the bottom ; the mother-liquor, rich in magnesium and calcium, is then allowed to go to waste. A similar process frequently takes place in nature, but the salts precipitated collect in the same basin in alternating layers.

In the Soda lakes near Ragtown, Nevada, a double carbonate of sodium and calcium, known as the mineral gaylussite, forms on the bottom, owing to natural concentration. When the waters are still farther evaporated, sodium sulphate and sodium carbonate are precipitated previous to the crystallization of common salt.

It has been found on concentrating sea-water that calcium carbonate is usually the first constituent to be precipitated. This salt is not always found when the waters of the ocean are analyzed, but may usually be detected in samples taken near shore. The vast quantity delivered to the ocean by rivers is soon eliminated by plants and animals and secreted in their tissues.

The succession of chemical precipitates formed in sea-water has been described by M. Dieulafait [1] as follows :

"First a very weak precipitation occurs of carbonate of lime (calcium carbonate), with a trace of strontium, and of hydrated sesquioxide of iron, mingled with a slight proportion of manganese. The water then continues to evaporate, but remains perfectly limpid, without forming any other deposit than the one I have mentioned, till it has lost 80 per cent of its original volume. It then begins to leave an abundant precipitate of perfectly crystallized sulphate of lime with two equivalents of water or gypsum, identical in geometrical form and chemical composition with that of the gypsum-beds. This deposit continues until the water has lost 8 per cent more of its original volume ; then all precipitation ceases till 2 per cent more of the original quantity of water has evaporated away. Then a new deposit begins, not of gypsum, but of chloride of sodium, or sea salt. . . . The deposition of pure or commercial salt continues till the volume of the water has been again reduced by one-half, when a precipitation of sulphate of magnesium begins to take place with it. This continues, the two salts being deposited in equal quantities, till only 3 per cent of the original quantity of water is left. Finally, when the water has been concentrated to 2 per cent, carnallite, or the double chloride of potassium and magnesium, is deposited. Spontaneous evaporation cannot go much further. The residual mother-water will not dry up at the ordinary temperature, even in the hottest regions of the globe ; its chief constituent is chloride of magnesium. A body of sea-water evaporated naturally will, then, leave a series of deposits in which we will find, as we dig down, the following minerals in order: deliquescent salts, including chiefly chloride of magnesium ; carnallite, or double chloride of potassium and magnesium ; mixed salts, including chloride of sodium and sulphate of magnesia ; sea-salt, mixed with sulphate of magnesia ; pure sea-salt ; pure gypsum ; weak deposits of carbonate of lime with sesquioxide of iron, etc."

[1] Popular Science Monthly, October, 1892.

In the natural evaporation of water in enclosed basins the succession will seldom be as regular as described above, for the reason that the process is apt to be interrupted by the addition of fresh supplies of water, and the succession begun anew, or else chemical changes initiated which will vary the results. In this connection it is to be noted also that changes of temperature, as from summer to winter, may modify the succession of salts deposited.

The separation of sodium sulphate, potassium chloride, and common salt from the mother liquor derived from the concentration of sea-water, by alternate evaporation and cooling, is the principle of Balard's well-known process largely used for obtaining salt from sea-water in the south of Europe. In Mesel's modification of this process, a low temperature is obtained artificially. When sea-water is concentrated until its specific gravity is 1.24 (28° of Beaume's hydrometer) it deposits about four-fifths of the common salt it originally contained; after adding 10 per cent of sea-water to the mother liquor remaining, it is passed through a refrigerating machine and its temperature lowered to $-18°$ C. The low temperature causes double decomposition to take place between the magnesium sulphate and the sodium chloride, sodium sulphate being deposited and the magnesium chloride remaining in solution.[1]

A process similar to that just described occurs in nature, as is shown by the precipitation of large quantities of sodium sulphate from the waters of Great Salt lake, during cold weather. This anticipation of Balard's process is noticed in advance in connection with other features of Great Salt lake.

The correspondence between the succession of salts formed by the evaporation of sea-water, and the succession found in many saline deposits deeply buried in the earth's crust, is of great interest and no doubt explains the genesis of some natural accumulations of this character. It is not always necessary, however, in seeking an explanation of the origin of beds of common salt, gypsum, etc., found in lenticular masses among stratified rocks, to assume that they were precipitated from isolated bodies of sea-water. On the contrary, the study of saline lakes has shown that similar deposits may result from the long concentration of ordinary river waters. So far as we are at present concerned, however, the process in either case is the same, since the waters of the ocean itself owe their salinity in a great degree to the concentration of the waters of streams.

[1] Report of Juries : International Exhibition, 1862, Class II, pp. 48–54.

Knowing the succession in which various salts are eliminated when waters are concentrated by evaporation, it becomes possible to determine in some instances, from the succession of salts discovered in a desiccated lake basin, what changes occurred in the life of the lake from which they were precipitated. In the case of Lake Lahontan, described in advance, this method has led to interesting conclusions. In a similar way, the chemical composition of a lake enables one to draw important inferences in reference to its past history. A lake in which the rarer elements found in tributary streams are abundant, must have undergone a long period of concentration, and formed deposits of the more common and less readily soluble salts. If a lake occupying an inclosed basin which has never overflowed, contains but a small percentage of the salts most common in the inflowing streams, it is evident that there must be some process by which such salts may be eliminated without being flooded out. Search should then be made for this new principle.

When lake waters are concentrated the first precipitates formed, as already seen, are ferric oxide and calcium carbonate. These substances are retained in solution mainly by reason of the presence of carbon dioxide, carbonic acid, in the water. As evaporation progresses and also when the waters are agitated, as in the breaking of waves on shore, the carbon dioxide escapes and the iron and lime previously held in solution are precipitated. The iron while in solution is in chemical combination with the carbon dioxide, forming ferric carbonate, when it loses its carbonic acid it is precipitated as ferric oxide. The lime in solution is believed to be in the form of the bicarbonate, and on losing carbon dioxide is precipitated as the carbonate.

It is a fact of geological interest that iron and lime held in solution may also be precipitated on account of the withdrawal of carbon dioxide through the agency of plant life. Low forms of vegetation, known as algae, thrive in the waters of both fresh and saline lakes and even in hot springs where the temperature approximates to that of boiling water. Through the vital action of this vegetation carbon dioxide is removed from water in much the same manner that higher forms of plant life eliminate it from the atmosphere. Carbon is assimilated and oxygen liberated. Iron on parting with its carbon dioxide unites with the liberated oxygen and is precipitated as ferric oxide.

It has recently been shown that large deposits of both calcareous tufa and silicious sinter are deposited through the agency of fresh water algae from the waters of hot springs in the Yellowstone Park. The silicia in such instances seems to be secreted by the plants as a part of their vital function, but the process is not well understood.[1]

The origin of oölitic sand, consisting of little spheres formed of concentric coats of calcium carbonate, along the shores of Great Salt lake, has been referred to an analogous process.[2]

Coral-like growths of calcareous tufa in some of the strongly alkaline lakes of the Great Basin are also thought to owe their origin to the agency of low forms of plant life.[3]

An important feature in this function of sub-aqueous plants, is that calcium carbonate may be precipitated from waters that are far below the point of saturation. In some instances precipitation is known to occur in this manner from water in which chemical tests fail to reveal more than a trace of calcium.

Ferric oxide is not known to be an important deposit in any of the lakes of North America, although found in abundance in many swamps. In Sweden, however, its precipitation from the water of fresh lakes is so abundant that it is of commercial value. The iron is carried into the lakes by streams, as a carbonate, and is precipitated on account of the loss of carbon dioxide, in part at least, through the agency of low forms of vegetative life. In some instances diatoms are thought to play an important part in secreting the iron.

With this brief sketch of the manner in which precipitates may be formed in lakes, let us turn to actual cases where the process is in operation. Of the considerable number of saline lakes of North America that have been studied, two are here selected as types. These are Great Salt lake, Utah, and Mono lake, California.

Great Salt lake, Utah. — This celebrated sea is situated in the eastern portion of the Great Basin near the west base of the Wasatch mountains. Its hydrographic basin has an area of 54,000 square miles, and is divided into two strongly contrasted portions. The eastern part is mountainous and contains peaks 12,000 feet in height above the sea, or 8000 feet above the lake. The western portion is composed of desert valleys but little elevated above the lake surface, and separated by narrow, abrupt, desert ranges rising from one to two thousand feet or more above adjacent valleys.

[1] W. H. Weed. "The Formation of Travertine and Silicious Sinter by the Vegetation of Hot Springs," 9th Annual Report, U. S. Geological Survey, 1887–88, pp. 613–676.

[2] A. Rothplitz. "On the Formation of Oölite," American Geologist, vol. 10, pp. 279–282.

[3] I. C. Russell. "A Reconnoissance in Central Washington," Bulletin No. 108, U. S. Geological Survey, pp. 94–95.

The elevation of the lake's surface varies somewhat during different years and from season to season, owing to climatic changes, and to the fact that the flow of the streams supplying it is interfered with for purposes of irrigation. Surveys made May 16, 1883, gave a surface level of 4218 feet above the sea.

Its area is also changeable. On a map made from surveys under the direction of Lieut. Stansbury, in 1850, it is represented as having an area of about 1750 square miles. A second map, made in connection with the Fortieth Parallel survey, in charge of Clarence King, in 1869, shows an area of 2170 square miles ; the increase in 19 years being 420 square miles, or 24 per cent. Its outlines when these surveys were made are shown in Plate 15.

At its highest observed stage in 1869, it had a maximum depth of 49 feet, and an average depth of approximately 19 feet. In 1850, the maximum depth was 36 feet, and the average about 13 feet. Since 1875, careful records of the fluctuations of level have been made and both annual and secular changes noted.[1] The annual high-water stage occurs in June, and is due to the melting of the snow on the Wasatch and Uintah mountains. The fluctuations embracing a series of years have not been found to be regular in their periods and are not coincident with observed climatic changes.

The shores of Great Salt lake are low except where a mountain uplift projects into it from the north, forming a rocky promontory, and for a short distance on its south shore where it touches the northern end of the Oquirrh mountains. Its surface is broken by several islands, of which two are short mountain ranges of the type so characteristic of the Great Basin. These rise more than a thousand feet above its surface and are rugged and precipitous. They stand like Nilometers in the saline waters, and on their sides are many horizontal lines marking former levels of the lake's surface. The highest of these scorings is about 1000 feet above the present water surface.

The scenery about this great lake of the Mormon land and in the encircling mountains is unusually fine, in spite of the aridity and the generally scant vegetation of the region. The sensation of great breadth that the lake inspires, together with the picturesque islands diversifying its surface, and the utter desolation of its shores, give it a hold on the fancy, and wakens one's sense of the artistically beautiful in a way that is unrivaled

[1] The records of these changes up to 1890, together with a discussion of their significance, is given by G. K. Gilbert, in Monograph No. 1, U. S. Geological Survey.

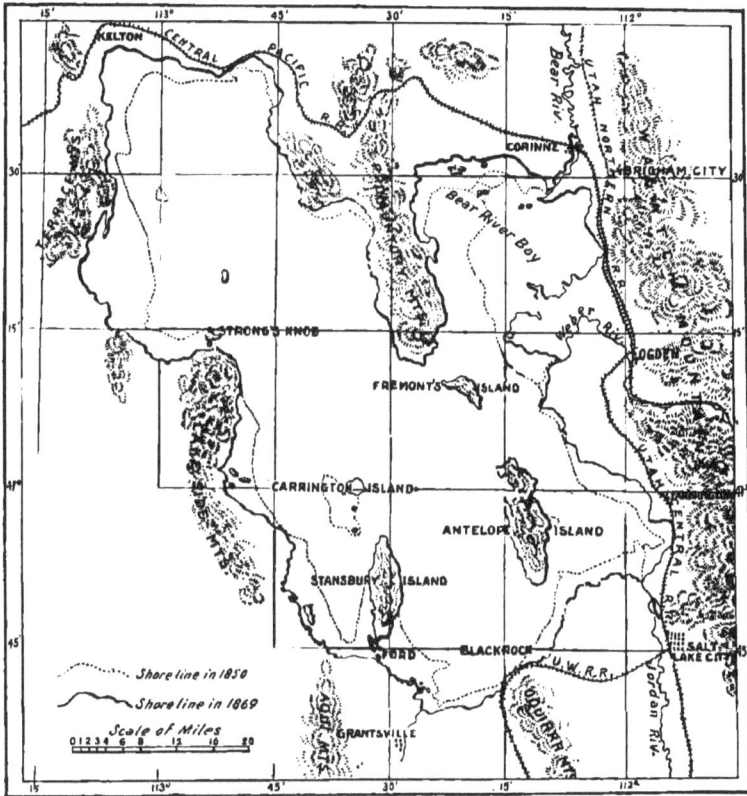

GREAT SALT LAKE, UTAH. (AFTER GILBERT.)

by any other lake of the Arid region. The unusually clear air of Utah, especially after the winter rains, renders distant mountains remarkably sharp and distinct, particularly when the sun is low in the sky and a strong sidelight brings the sharp serrate crests into bold relief and reveals a richness of sculpturing that was before unseen. At such time the colors on the broad deserts, and amid the purple hills and mountains, are more wonderful than artists have ever painted, and exceed anything of the kind witnessed by the dweller of regions where the atmosphere is moist and the native tints of the rock concealed by vegetation. The hills of New England when arrayed in all the gorgeous panoply of autumnal foliage are not more striking than the desert ranges of Utah when ablaze with the reflected glories of the sunset sky. The rich, native colors of the naked rocks are then kindled into glowing fires, and each cañon and rocky gorge is filled with liquid purple, beside which even the Imperial dyes would be dull and lusterless. At such times the glories of the hills are mirrored in the dense water of the lake ; their duplicate forms appearing in sharp relief on the paler tints of the reflected sky. As the sun sinks behind the far-off mountains, range after range fades through innumerable shades of purple and violet until only their highest battlements catch the fading glory. The lingering twilight brings softer and more mysterious beauties. Ranges and peaks that were concealed by the glare of the noon-day sun, start into life. Forms that were before unnoticed, people the distant plain, like a shadowy encampment. At last each remote mountain crest appears as a delicate silhouette, in which all details are lost, drawn in the softest of violet tints on the fading yellow of the sky.

To one who only beholds the desert land bordering Great Salt lake in the full glare of the unclouded summer sun, when the peculiar desert haze shrouds the landscape and the strange mirage distorts the outline of the hills, the scenery will no doubt be uninteresting and perhaps even repellent. But let him wait until the cool breath from the mountains steals out on the plain and the light becomes less intense, and a transformation will be witnessed that will fill his heart with wonder.

The saline and alkaline shores of Great Salt lake are either naked mud plains, frequently white with drifting salts, or scantily clothed with desert shrubs. The absence of conspicuous flowers is frequently relieved by broad areas covered with a peculiar plant, known as *Salicornia*, which flourishes by the side of this Dead Sea of the West, where all other vegetation perishes. The *Salicornia* grows in fleshy stems, without leaves, and looks not unlike branching coral. It is of many shades of red, pink, and yellow, thus still further increasing its resemblance to groves of living coral. The white, alkaline desert is frequently tinted by this strange plant until it glows like a field of Alpine flowers. There are many other interesting features to be noted by the visitor to the great desert-lake of Utah, but its physical and chemical history claims our attention at this time rather than its artistic setting.

The streams flowing to the lake rise in the high mountains to the east and are clear and limpid, and of such purity that only chemical tests reveal the presence of the mineral matter they have dissolved from the rocks and soils. Several of these streams are truly rivers in volume, as well as in name, and send a never-ceasing flood to the lake. Their combined volumes average throughout the year about 10,000 cubic feet per second.[1]

There are a number of fissure springs about the lake, or rising beneath its surface. In some instances these are hot and contain more saline matter in solution than is usually found in surface streams. These contribute a considerable quantity of the saline matter found in the waters of the lake, but it is believed that the amount thus derived is less than that furnished by streams from the mountains. This conclusion rests on incomplete data, however, as neither the volume nor the composition of all the springs is known. None of the springs supplying the lake, with a single known exception, of small volumes, are markedly saline. The salts they contain are acquired largely during the upward passage of the water through the sediment of former lakes and their influence on the chemistry of the present lake is more important than in the case of any other lake in the same region. It is safe to conclude, however, that the combined volumes of the streams and springs now tributary to the lake, if not concentrated by evaporation, would form a water body in which no trace of saline matter would be apparent to the taste.

Analyses of the waters of Bear river, of Utah lake, from which the Jordan flows, and of City creek, one of the numerous streams from the west slope of the Wasatch mountains, give an average of about 0.2446 part per thousand of mineral matter in solution. This may be taken as the average composition of the surface stream flowing to the lake. As will be noticed on referring to the average composition of normal rivers previously given, the mineral matter in these streams is nearly double the amount carried in the same volume of water by streams

[1] G. K. Gilbert. "Lands of the Arid Region," Washington, 1870, p. 72.

in more humid regions.   This is due, in a measure, to the active evaporation that takes place from them and from the lakes on their courses.

The waters of Great Salt lake have been analyzed at six different
times.   The results of these several analyses are widely at variance on
account of fluctuations in the volume of the lake.   The dates at which the
various samples analyzed were collected and the total solids found in 1000
parts of water are here given : [1]

Date . . .	1850	summer 1869	Aug. 1873	Dec. 1885	Aug. 1889	Aug. 1892 [2]
Specific gravity	1.170	1.111	1.102	1.122	1.157	1.156
Parts in 1000	224.2	148.2	136.7	167.2	195.5	205.1

Since the accompanying table of analysis of lake waters was compiled,
my attention has been directed to the analysis given below, which in
several ways is the most complete and satisfactory that has been published.

ANALYSIS OF A SAMPLE OF THE WATER OF GREAT SALT LAKE.   COLLECTED
AUGUST 9, 1892.[2]

By E. WALLER.

[Expressed in Grams in a Liter. Specific Gravity, 1.156.]

ELEMENTS AND RADICALS.		PROBABLE COMBINATION.	
Na	75.825	$NaCl$	192.860
K	3.925	K_2SO_4	8.756
Li	0.021	Li_2SO_4	0.166
Mg	4.844	$MgCl_2$	15.044
Ca	2.424	$MgSO_4$	5.216
Cl	128.278	$CaSO_4$	8.240
SO_3	12.522	Fe_2O_3 and Al_2O_3	0.004
O in sulphates	2.494	SiO_2	0.018
Fe_2O_3 and Al_2O_3	0.004	Surplus SO_3	0.051
SiO_2	0.018	Total	230.355
BO_2O_3	Trace	Total solids by evaporation	238.12
Br [3]	Faint trace	" " [duplicate] . . .	237.925

The average composition of the combined spring and stream waters
tributary to the lake cannot be stated with accuracy, but judging from

[1] A compilation of various analyses of the water of Great Salt Lake and a discussion concerning them, is given by G. K. Gilbert, Monograph No. 1, U. S. Geological Survey.

[2] School of Mines [Columbia College] Quarterly, vol. 14, 1892, p. 58.

[3] A later determination showed about 0.01 gram of Br. in a liter.

such observations as bear on the question, it seems safe to assume that their mingled waters would contain less than double the percentage of saline matter found in the surface streams. The assumption that the combined spring and stream waters would contain about 0.3 part in a thousand, or three one-hundredths of one per cent of total solids in solution, seems as close an approximation as can now be reached.

The waters of the lake during recent low stages have become nearly saturated with sodium chloride and sodium sulphate, and under certain conditions these salts are precipitated. The point of saturation for calcium carbonate is passed, and this salt is precipitated probably as rapidly as it is received. The waters are not rich in the compounds of bromine, boron, lithium, and iodine, which frequently occur in "mother-liquors," remaining when the more common salts have been eliminated by long concentration, and hence indicating the old age of a lake containing them. The recent analysis by Waller, however, shows these rarer elements to be present in somewhat larger quantities than was previously supposed.

The length of time that would be required to charge Great Salt lake with the common salt it contains, under the present conditions, is estimated by Mr. Gilbert at about 25,000 years.

The quantity of sodium chloride, or common salt, held in the water of the lake is estimated at 400 million tons, and the sodium sulphate at 30 million tons. These figures indicate the commercial importance of this great reservoir of brine. The separation of the common salt has already led to a considerable industry, as from 20 to 40 thousand tons have been gathered yearly for a considerable period. The most extended and best conducted of these operations are carried on by the Inland Salt Company at the southern end of the lake. Evaporating vats covering more than one thousand acres have been constructed, and are supplied by pumps which deliver 14,000 gallons of lake water per minute. Pumping is continued through May, June and July, and the salt is ready for gathering in August. During midsummer the amount of water evaporated is 8,400,000 gallons daily. The yield of salt is at the rate of 150 tons per inch of water per acre. An average season's yield is a layer of salt about seven inches thick, which would be precipitated from forty-nine inches of water. The facilities for this industry may be judged by the fact that coarse salt packed on cars ready for shipping, is sold at the works for one dollar per ton. The mother-liquor is allowed to go to waste, but it is to be expected that sodium sulphate and other salts contained in it will be utilized in the near future.

Along the margin of Great Salt lake, where the water is only a few inches deep, it becomes so concentrated by evaporation that common salt crystallizes and forms a brilliant white layer on the bottom. In fording an arm of the lake about a mile broad, in order to reach Stansbury island, the writer, in 1880, found a crust of salt forming a glistening pavement strong enough to support a horse and rider, but occasionally it would give way and lead to uncomfortable flounderings in the black mud beneath.

The solubility of sodium sulphate is controlled largely by temperature. In Great Salt lake in summer it is all dissolved and the waters are clear, but as cold weather approaches it separates and renders the waters opalescent and somewhat milky in color. In the depth of winter, when the temperature falls below zero of the Fahrenheit scale, as it does at times for days together, this salt separates in great abundance and is thrown ashore by the waves in hundreds of tons, forming a slush-like mass on the beach looking like soft snow. On such occasions it can be gathered in practically unlimited quantities, but is soon re-dissolved when the temperature rises.

The brine of the lake is so concentrated that fish cannot live in it, but it furnishes a congenial home for small crustaceans known as brine shrimps (*Artemia*) and for the larvae of dipterous insects. These are abundant at certain seasons, but not in such vast numbers as in some of the more alkaline lakes on the west side of the Great Basin. It has been stated that the vast numbers of crustaceans and of larvae in these waters are due to the fact that there are no fishes or other animals in the lakes that could prey upon them; aquatic birds, however, feed upon them in great numbers, but still they swarm in countless myriads. Their food seems to be minute algae of which several species have been described.

As shown by the analysis given above, the principal salt in Great Salt Lake is sodium chloride. In the second example of the saline lakes described below the characteristic ingredients are sodium carbonate and sodium sulphate. Great Salt lake may be said to be a *salt lake* in distinction from a number of water bodies situated especially on the west side of the Great Basin, which may with propriety be designated as *alkaline lakes*.

Mono lake, California. — This lake, selected as the type of a series of strongly alkaline water-bodies in the desert basins of the Arid region, is situated in south-eastern California, within a few miles of the Nevada boundary. It lies at the immediate eastern base of the Sierra Nevada, from which it receives practically all of its water supply, and occupies one of the minor basins composing the great area of interior drainage known as the Great Basin. Its position on the west side of the Great Basin and at the base of the great fault scarp forming the precipitous eastern slope of the Sierra Nevada, is similar to the situation of _Great Salt Lake on the east side of the same broad desert area, and at the west base of the magnificent fault scarp forming the abrupt western face of the Wasatch range. Mono lake, like many other enclosed water bodies of the Arid region, is of ancient lineage, as is shown by numerous beach lines, carved by former water bodies, on the inner slopes of this valley. The highest of these lines is from 670 to 680 feet above the present water surface.

The hydrographic basin of Mono lake has an area of nearly 7000 square miles, and, as in the case of the region draining to Great Salt lake, is divided into two strongly contrasted portions. The southwestern part is mountainous and rugged, and bristles with serrate peaks that rise over six thousand feet above the lake's surface. On the mountains the snow-fall is abundant, and several small glaciers exist in the higher valleys. The eastern portion of the drainage basin is comparatively low, and is arid and desert-like in character. Little rain falls on this portion of the basin, and there are no perennial streams. Only occasionally is there sufficient precipitation to produce a surface drainage, and normally the rain water and the water produced from the melting of the light winter snows, is absorbed at once by the thirsty soil or returned to the atmosphere by evaporation.

To gain a comprehensive idea of the geography of the interesting region about Mono lake, one should climb some commanding summit on the High Sierras, on its southwestern border, and study the magnificent panorama spread out at his feet. Let the reader come with me to the summit of Mt. Dana, named in honor of the venerable J. D. Dana, one of the most prominent peaks overlooking Mono lake, and I will endeavor to point out some of the more interesting features of the land we are studying.

The summit we have reached is nearly 13,000 feet above the sea. The only neighboring mountains exceeding it in altitude are Mt. Lyell and Mt. Ritter, which rise with dazzling whiteness against the southern sky. From our station the entire Mono basin is in view, and much of its history can be read as from a printed page. We are standing on one of the

PLATE 16.

MonoCraters   Mt Ritter   Parker Cañon   Bloody Cañon   Mt Gibbs (Gibbs) Cañon   Mt Dana   Leevining Cañon   Paoha Island   Mt Warren   Negit Island   Lundy Cañon   Dunderberg Pk.

THE HIGH SIERRA FROM THE NORTH SHORE OF MONO LAKE.

highest points on the rim of a sharply defined hydrographic basin. The
drainage from all directions tends towards the center and forms a lake
from which the waters escape only by evaporation. We can trace nearly
the entire boundary line of the basin, for the reason that the slopes are so
plainly marked and the crest lines so sharply drawn, that there is no doubt
as to the direction that surface water would take. The courses of the
swift, bright stream descending the mountain can be followed from their
sources in melting snow-fields, down through deep cañons to where they
enter the lake. On the desert side of the basin, however, there are no
streams, and but indefinite traces of the dry beds of former water-courses.
There is no notch in the rim of the basin to suggest a former outlet. The
only possible point of discharge for the waters when the ancient beaches
scoring the inner slopes of the valley were formed, is far to the north, and
concealed from view. Apparently at our feet, but in reality a mile in
vertical descent below, lies the lake, a silent and motionless plain of blue.
Should the wind chance to be strong in the valley, however, its surface
would be ruffled, the flash of breaking waves would reach the eye, and
long lines of froth would streak its surface. At such times a broad fringe
of snowy foam, produced by the churning of the alkaline waters, encircles
the shores and renders their outlines unusually distinct. Apparently
floating on the surface of the lake, there are two conspicuous islands, the
forms of which show that they are of volcanic origin. That these craters
were built since the encircling waters fell below their level, is shown
by their unbroken contours and by the absence of terraces on their outer
slopes.

Beyond the lake the brown and barren land seems low and featureless, because of the elevation of our point of view. We can see far beyond the limits of the drainage basin, in which we are now specially
interested, and distinguish many of the desert ranges of Nevada rising
above the purple haze enshrouding their bases and obscuring the lifeless
lands between. The highest of these distant summits, which appears like
a spectral mountain floating in the sky, is even higher than the peak
on which we stand, but its naked sides are scorched to a cinder-like
redness by the desert heat, and no silvery stream can be detected in
the wild gorges scoring its flanks. Its summit is seldom cloud-capped,
and only in the depth of winter is its ruggedness concealed by a
mantle of snow.

To the right of the lake is a long range of craters built of fragments
of volcanic rock thrown out during many violent eruptions, and now

forming conical piles with gracefully sweeping outlines. Several of these now silent volcanoes rival Vesuvius in height and beauty, but from our elevated stations we can look down upon the depressions in their summits, and the entire range, although two miles in length, with peaks rising three thousand feet above the lake that bathes its feet, is but a minor feature in the extended landscape.

Northwest and southeast from the summit of Mt. Dana the crest line of the Sierras is marked by mountain after mountain as far as the eye can reach. Turning south we have in view a fine example, though not the very finest, of the wild and rugged High Sierras. At the western base of the Mount Dana there is a deep, picturesque valley, dotted with lakes and traced by gleaming streams. Like nearly all of the more pronounced depressions in the High Sierras, this valley owes its origin principally to stream erosion, and is a relic of an ancient drainage system, but has been enlarged and its minor features modified by ice abrasion. At one time it was occupied by a glacier, which formed a part of a great system of ice fields that covered all of the High Sierras and sent many ice streams both to the east and west, through precipitous gorges to the valleys below.

The rocks forming the nearer slopes as one looks toward the more rugged portion of the range are of varied character and rich in color ; but farther within the heart of the mountains the monotonous gray coloring of granite is but partially concealed by the scanty forests in the cañons and valleys, or by the mosses and lichens on the higher summits. Near at hand, but across a deep intervening valley, rises Mt. Conness, bare, rugged, and grand. Twelve miles to the south, across a fragment of deeply eroded table-land, named the Kuna crest, are the spire-like peaks of Mts. Lyell and Ritter. Throughout the year their summits are white with snow, and small glaciers can be distinguished in the folds of their rugged sides. Returning from this vision of wild magnificence, the eye rests upon a scene humbler in its charms but not less pleasing. Between the naked crags forming the summit from which we have gained our commanding view, and the highest limit of the pines, all twisted and deformed from unequal struggles with wind and drifting snow, there is a belt of rugged precipices and weather-beaten rocks that at certain seasons are bright with lichens and fringed with the purple and gold of alpine blossoms. These charming decorations on the mountain's brow flourish with rank luxuriance in every cranny and cleft, and not infrequently are in such rich profusion that an entire summit-peak is tinted by them as with a twilight glow.   In these elevated regions May-day is a festival of late summer, but it brings with it a multitude of charms that are unknown to dwellers in the world below.

The mountains hold out innumerable charms to detain us, but we must descend in our fireside journey, and learn more of the strange lake, the setting of which was revealed from our station on the mountain top. Our downward journey is through a deep gorge with nearly vertical walls; in its bottom a swift, clear stream plunges from ledge to ledge, and rushes through rocky chasms with a roar that never allows the echoes of the cliffs a moment's pause.   This pure stream of cold, delicious water reveals the character of many creeks and rivulets that are rushing down the mountain side to the ever-thirsty valley below.

A few springs add their waters to the supply from the mountains, but none of them are saline, and their united volume is far less than the volume of any one of half-a-dozen of the mountain torrents pouring into Mono lake.   The present density of the lake water is the result of the long concentration by evaporation of the supply from the mountains.

The area of Mono lake in the summer of 1883, was 87 square miles, but varies with the seasons and also from year to year.   As may be learned from the accompanying map, its north and south axis measures 11, and its east and west axis 14 miles.   Its surface is broken by two volcanic islands and by numerous crags, some of which are remnants of islands now nearly eroded away, while others are formed of calcareous deposits precipitated about submerged springs.   The soundings given on the map, show that its maximum depth is 152 feet, and the mean depth about 61 or 62 feet.   Its elevation above the sea, when surveyed in 1885, was 6380 feet.

In Pleistocene times, when great glaciers descended from the High Sierras and were prolonged several miles into the valley, the ratio between inflow and evaporation was changed, and the lake rose, but never sufficiently to discover an outlet.   During the time of its greatest expansion, it had an area of 316 square miles, and formed an unbroken water surface 28 miles long from north to south, and 18 miles broad.   Its maximum depth was then over 800 feet.

The facts of greatest interest in connection with Mono lake are to be found in its chemical history.   As shown in the analysis of its waters given on page 72, it is strongly impregnated with sodium and with carbonic and sulphuric acids.   The most probable combination of these and other substances present in the waters is given below :

HYPOTHETICAL COMPOSITION OF THE WATER OF MONO LAKE.

BY T. M. CHATARD.[1]

CONSTITUENTS.	GRAMS IN A LITER.	PER CENT OF TOTAL SOLIDS.
Silica, SiO_2	0.0700	0.13
Aluminum and ferric oxide $(Al_2Fe_2)O_3$	0.0030	0.005
Calcium carbonate, $CaCO_3$ . . .	0.0050	0.09
Magnesium carbonate, $MgCO_3$ . .	0.1928	0.36
Sodium borate, $Na_4B_4O_7$ . . .	0.2071	0.39
Potassium chloride, KCl . . .	1.8365	3.44
Sodium chloride, $NaCl$ . . .	18.5033	34.60
Sodium sulphate, Na_2SO_4 . . .	9.8690	18.45
Sodium carbonate, Na_2CO_3 . . .	18.3556	34.33
Sodium bicarbonate, $NaHCO_3$ . .	4.3856	8.20
[Specific gravity, 1.045.]	53.4729	100.00

As may be seen in the above table, sodium carbonate and bicarbonate form 42.53 per cent of the total salts held in solution. The total quantity of these salts contained in the lake is estimated at 92 million tons, the total saline content being 245 million tons.

Owing to the cost of transportation and the high price of labor, this brine is not now utilized, but it forms a reservoir that may be drawn upon in the future. The waters of Owens lake, situated a hundred miles south of Mono lake, where the commercial conditions are somewhat more favorable, is already the basis of a large soda industry. Two small lakes on the Carson desert, known as the Ragtown ponds, or Soda lakes, also furnish large quantities of sodium carbonate and bicarbonate. There are also several other lakes of the same general character in the western part of the Great Basin which have not yet been found of economic importance. One of the most promising of these, from a commercial point of view, is Soap lake, in the State of Washington.

The great abundance of sodium carbonate and bicarbonate in Owens, Mono, and other lakes on the west side of the Great Basin, in contrast with the amount of these salts in the brine of Great Salt lake and of other similar water bodies on the east side of the Great Basin, is due mainly to differences in the character of the rocks of the two regions. The mountains on the west are largely formed of volcanic rocks, and yield alkaline

[1] Amer. Jour. Sci., 3d Ser., vol. 36, 1888, p. 149.

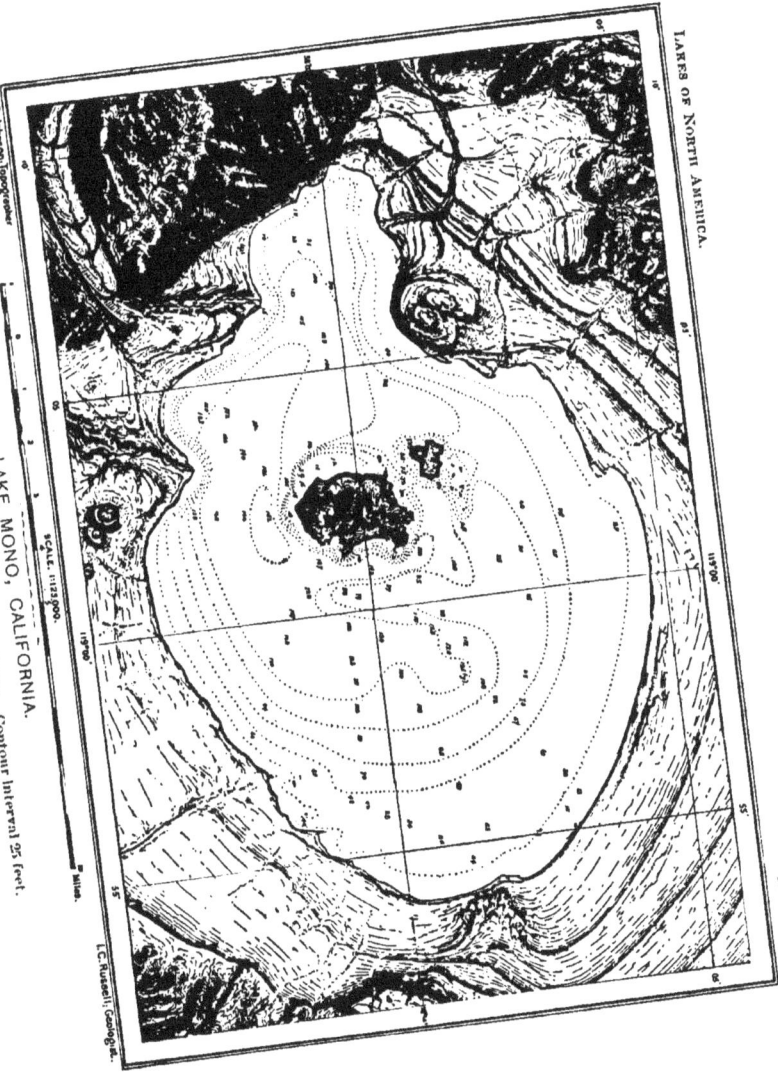

LAKE MONO, CALIFORNIA.

Subaqueous contours approximate. Contour interval 25 feet.

Soundings in feet.

W. D. Johnson, Topographer.

SCALE, 1:125,000.

I. C. Russell, Geologist.

PLATE 17.

salts to the waters flowing over them or percolating through their inter-
stices ; while the rocks of the eastern area are largely sedimentary in
origin, and supply sodium chloride in excess of sodium carbonate.

The chemical history of the lakes of the Arid region is not only an
interesting and attractive study, but one of great economic importance,
as they hold an almost unlimited supply of common salt, and of sodium
carbonate and bicarbonate, sodium sulphate, and other salts in less abundance.   This supply is still farther augmented by the deposits of former
lakes now evaporated to dryness.   The salts precipitated from these extinct lakes, in some instances, whiten the surfaces of desert valleys, but
more frequently they are buried beneath or absorbed in the clays forming
the smooth plains left by the evaporation of playa lakes.

The importance of the lakes of the Arid region to those interested in
salt and alkali industries is so great that the table on page 72 has been
inserted to show the comparative values of the brines thus far analyzed.
More detailed information in this connection may be found in the publications cited below.[1]

[1] G. K. Gilbert, "Lake Bonneville," U. S. Geol. Surv., Monograph No. 1. — I. C. Russell,
"Lake Lahontan," U. S. Geol. Surv., Monograph No. 11. — I. C. Russell, "Lake Mono,"
U. S. Geol. Surv., 8th Ann. Rep., 1886–87, pp. 287–299. — I. C. Russell, "Reconnoissance in
Washington," U. S. Geol. Surv., Bulletin No. 108. — T. M. Chatard, "Natural Soda," U. S.
Geol. Surv., Bulletin No. 60. — T. M. Chatard, "Analyses of the Water of Some American
Alkaline Lakes," Am. Jour. Sci., 3d Ses., vol. 36, 1888, pp. 146–150. — T. M. Chatard,
"Urano," Am. Jour. Sci., 3d Ses., vol. 38, 1889, pp. 59–66. — J. E. Talmage, "The waters
of Great Salt Lake," Science, vol. 14, 1889, pp. 444–446. — E. Waller, "Analysis of the
Water of Great Salt Lake," School of Mines [Columbia College] Quarterly, vol. 14, 1892,
pp. 56–61.

CHAPTER V.

THE LIFE HISTORIES OF LAKES.

LAKES, like many other features of the earth's surface, as stated in our introductory chapter, have their periods of growth, adolescence, maturity, decadence, and old age leading to extinction.

The lives of most lakes are so long that human records cover only a small portion of their histories, hence their growth and decadence can seldom be traced by observing a single individual. By studying many examples, however, in various stages of development and decline, we are enabled to obtain separate links in the chain of their existence, and may determine, at least in outline, the general course that they run. By having the theoretical history of a normal lake in mind, one is enabled to determine the period of life attained by any special example that may be studied.

The histories of all lakes are far from uniform. There are various accidents, as they may be termed, which introduce new conditions, and may renew their youth or hasten their decline. In general, lakes may be grouped in two great classes, in each of which the rôle they play is in the main the same. The differences in the lives of these two classes depend mainly on climatic conditions, and have been noticed in describing fresh lakes and terrestrial saline lakes. The destiny of a lake born beneath humid skies runs in a somewhat definitely prescribed channel and departs in a marked way from the more varied life of a lake originating in an arid region. The general outline of the history of each of the two classes referred to is briefly as follows:

Lakes of Humid Regions. — The normal lakes of humid regions are comparatively short-lived. The streams tributary to them bring in sediments which tend to fill their basins, to these are added the débris of water-loving plants and the hard parts of animals, and at the same time the streams flowing from them tend to cut down their outlets and drain them at lower and lower levels. Two processes thus conspire to diminish their volumes and shorten their existence. The deposition of sediment on their bottoms usually leads to their extinction more quickly than the lowering of their outlets, for the reason that while incoming streams are frequently turbid and heavy with sediment, the outgoing waters are clear and therefore have but little power to erode. The clear outflowing waters deepen their channels by the slow process of chemical solution, but when the rocks over which they pass are soft and incoherent, they may soon become recharged with sediment and make rapid progress in deepening their channels and in draining the basin above. The lives of various lakes may differ in length and have minor variations according to local conditions, but the main features in their histories will conform to the same general outline.

The filling of lake basins by sediment frequently progresses more rapidly than at first might be supposed. In some instances its rate may be observed from year to year, and attracts the attention of even the casual observer. In countries that have been long inhabited, there is sometimes historical evidence of the rate at which the boundaries of lakes have contracted. At the head of Lake Geneva, Switzerland, for example, the Rhone is bringing in large quantities of silt derived from the glaciers on its head waters, and a low grade delta is being extended into the lake. As stated by Lyell,[1] the town of Port Vallais (Portus Valesæ of the Romans) once situated at the water's edge, is now more than a mile-and-a-half inland, this extension of the shore having been made in about eight centuries.

The decrease in the capacity of lake basins, in ordinary cases, goes on so much more rapidly from filling than from the lowering of their outlets, that it is the destiny of most lakes situated in humid regions to become exterminated mainly by sedimentation. By this process their basins are transformed into alluvial plains, through which wander the streams that were tributary to the antecedent lakes. These streams being no longer robbed of the material they carry in suspension, are enabled to attack their channels below the former lakes with energy, and to deepen and broaden them. The grade of the streams through the alluvial plain, marking the former site of a lake, is increased, and the removal of the soft lakebeds progresses as the channel below is deepened. Streams flow through alluvial plains with slackened speed, and form winding channels, and swing from side to side of their valleys, thus reducing the general level. The load previously deposited in the basin is again taken up and the deferred task of transporting it to the sea is resumed. Former lake-

[1] Principles of Geology, 11th edition, 1873, vol. 1, p. 413.

basins thus become terraced valleys, with streams winding through them
in broad curves, and in civilized regions afford rich farming lands and
charming sites for towns and cities.

At a later period, if some outside influence does not change the course
of history, the alluvial deposits are dissected to the bottom, the terraces
of soft material are removed, and all records of the once beautiful lake
may be lost. This transformation may require tens of thousands of years
for its completion, yet the end is inevitable. The various stages in this
general history might be illustrated by an abundance of examples. Thousands of lakes in the formerly glaciated region of northeastern America
still retain the freshness of youth, and their nearly level bottoms may be
considered as unborn lacustral plains. The terraced borders of Lake Champlain, and of the Laurentian lakes, mark the former extent of water
bodies that have passed the youthful stage. Many terraced valleys in the
Cordilleras record the former presence of lakes in basins that are now
completely drained. In other localities, as in the "Parks" of Colorado,
no terraces may be distinguished, but vestiges of lacustral sediment still
floor their bottoms. Many valleys in the same region drain through
narrow stream-cut gorges, but all other evidence of their having been
formerly water-filled has vanished. The time required for these muta-
tions is vast when reckoned in years, but to the geologist they are
transient phases in the topographic development of the land.

The even course of history, outlined above, may be varied somewhat,
as when the outflowing stream is rapid and especially when falls occur in
its course. Waterfalls are formed especially where streams flow over
nearly horizontal strata where a hard surface layer rests upon shales or
other easily eroded beds, as is typically illustrated at the Falls of Niagara.
The undermining of the hard capping layer is effected by the removal of
the soft beds beneath, and blocks from the brink of the precipice fall to
the pool below and assist the swirling water to deepen a basin. A fall
thus cuts back the ledge over which it plunges with comparative rapidity, — in the case of Niagara the rate of recession is from 4 to 6 feet per
year, — and may lead to the drainage of a lake before its basin has been
deeply filled with sediment. The succession of the principal events in the
history of a valley may thus be hastened, but the ultimate results will be
essentially the same.

Many small lakes, especially in forested countries, where the surface
waters filter through layers of vegetable débris before gathering into rills
and brooks, are filled mainly by organic agencies. Water plants, and

especially *Sphagnum* or peat moss, grow about their shores, and extending outward, form a thick mat of intertwined roots and stems that float on the surface. The finer waste from this sheet of floating verdure falls to the bottom and forms a peaty stratum. To this layer contributions are made by other aquatic vegetation, as the lilies, reeds, rushes, and many beautiful sub-aquatic plants. It also receives the trunks of trees falling from the shore. The small lakes of the prairie region especially, are frequently transformed in this manner into beautiful fields of wild rice. In the central part of moss-encircled lakes, practically no mechanical sediments are deposited, but mollusks, crustaceans, and fishes may there find a well sheltered home and thrive in such abundance that the bottom soon becomes covered with their remains. Microscopic forms also inhabit the water and their siliceous cases frequently accumulate so as to form thick layers, known as diatomaceous earth. A continuation of this process under favorable conditions leads to the rapid extinction of small lakes. The open waters are converted into bogs and swamps, on which forest trees encroach and still farther assist in the transformation. When these deposits of organic matter are drained, they frequently furnish rich garden lands. The lakes exterminated by this organic process in the drift-covered portion of North America, can only be estimated in tens of thousands, and probably equal in number the lakes still remaining.

Lakes of Arid Regions. — On every continent there are broad areas where the skies are without a storm cloud for many months each year and the air is dry and hot in all but the winter season. The lakes in these desert regions have a different general history from their sisters whose banks are fringed with green vegetation and overshadowed by forests. Where the rainfall is small and evaporation active, the lives of lakes depend on delicate adjustments of climatic conditions. As the barometer rises and falls in harmony with changes in atmospheric pressure, so enclosed lakes fluctuate in sympathy with changes in humidity or in temperature. The ephemeral lives of playa lakes have already been described, but the larger lakes of arid regions, although subject to many fluctuations, may have a longer span of existence than lakes of corresponding size and similar topographic environment in humid regions. As enclosed lakes do not overflow, there is no loss of area owing to the lowering of outlet. Tributary streams bring in material both in solution and in suspension, all of which is left as evaporation progresses, and tends to fill their basins, but the volume of their waters is not directly diminished by this process.

As their basins are filled, however, the waters expand and offer a greater surface to the atmosphere, thus promoting evaporation. A continuance of this process results in so enlarging the water surface that in time evaporation equals the supply and the water body passes to the condition of a playa lake. Sedimentation may raise the water surface so that an outlet is found before the playa stage is reached, thus transferring an enclosed and saline lake to the class normal to humid regions, already considered.

The existence of lakes in countries where there is a close adjustment between precipitation and evaporation, is also controlled largely by topographic conditions. It may be said that this is the primary condition that determines whether lakes shall exist in arid regions or not. This is true, if we consider the origin of the lake basins, and also important if the existence of lakes in ready-formed basins is discussed, since the topography has a direct and frequently controlling influence on rainfall and on evaporation. The influence of topography is also marked in determining the ratio of the area of a basin to the area of the lake in its lowest depression. The hydrographic basins of enclosed lakes as a rule are large in reference to their water surfaces, when compared with the ratio of catchment areas to lake areas in humid regions. Any change tending to diminish the area tributary to an enclosed lake, as the sapping of the head waters of its tributary streams, would have a marked influence on its history.

Episodes of another character also occur in the lives of enclosed lakes. The salts slowly accumulated in them may not only be flooded out by overflow consequent on changes in topography, or on an increase in rainfall, or on a decrease in evaporation, but may be eliminated by reason of a reverse change in the ratio of inflow to evaporation. A decrease in humidity or an increase in evaporation, or what is probably more frequent, a combination of these two processes, may reduce a lake to the playa stage. When this occurs, its salts will be precipitated and may become buried or absorbed by sediment, so that when a new lease of life is granted and the waters expand and form a perennial lake, they are fresh, or essentially so, and start anew in the process of concentration. Still other changes that beset the lives of enclosed lakes might be enumerated, to show that they are subject to greater vicissitudes than their sister lakes in more favored lands.

When the lakes of arid regions become extinct, either by reason of evaporation or sedimentation, the evidence of their former existence remains inscribed on the inner slopes of their basins or concealed in the strata deposited over their bottoms. These records as a rule are much more lasting than those left by lakes in humid lands, for the reason that the climatic conditions are less destructive. The terraces and embankments of gravel left by lakes in desert valleys are especially permanent topographic features, as the scanty rain that falls on them is absorbed and allowed to percolate slowly through them, instead of flowing down their surfaces so as to erode. The sediments deposited in enclosed basins are also protected from destruction, as they cannot be removed by streams until some change inaugurates free drainage to the sea or to some lower basin. A continuation of aridity in a desiccated lake basin, results normally in the burial of the lacustral sediments beneath subaërial deposits, thus again insuring their preservation. To follow this subject farther would lead to a comparative study of the processes of erosion in arid and in humid regions, which is beyond the scope of the present essay.

It will be seen from what has been presented above with reference to the normal course of the lives of lakes, that in spite of the many variations they present, the seeds of death are planted at their birth, and they are destined, sooner or later, to pass away and give place to other conditions.

Interruptions of the even tenor of the lives of lakes, in both arid and humid regions, such as the effects of upheaval and depression of the earth's crust, earthquakes and volcanic eruptions, might be considered, but these abnormal incidents, like the accidents in human lives, cannot be foretold, and apply to individuals rather than to classes.

CHAPTER VI.

STUDIES OF SPECIAL LACUSTRAL HISTORY.

It will appear to the reader of the preceding chapter that not only are lakes ephemeral features of the earth's surface, but even the changes they make in the topography of their shores, although perhaps engraved in solid rock, are of short duration in comparison with the length of the eras into which the earth's history has been subdivided. The lakes of Pleistocene times, however, left records which in many instances are still legible, and form a connection between historical and the most recent geological times.

As examples of extinct lakes whose histories are still clearly legible, a brief account will be given of former water bodies of the Laurentian basin, and in the region now draining to Lake Winnepeg, where the climate is humid, and of two formerly extensive lakes of the Arid region.

PLEISTOCENE LAKES OF THE LAURENTIAN BASIN.

Long curving ridges of gravel having the appearance of great railroad embankments, following the general trend of the shores of lakes Ontario and Erie, but usually at a distance of several miles from their present borders, were noticed at an early day in the settlement of New York, Ohio, and Ontario, and correctly interpreted as being the records of previous high-water stages of the lakes they encircle. These ridges became highways of travel as civilization advanced, and gave origin to the term "ridge road" still to be seen on local maps of the region referred to. These ridges and other associated records have claimed the attention of geologists and others and have been made the subject of special inquiry. The territory traversed by them is so extensive, however, that their study is still far from complete.

The ancient beaches about lakes Ontario and Erie have been followed and studied, especially by G. K. Gilbert, in New York and Ohio, and by J. W. Spencer, in Canada. The records of former water levels north of Lake Superior from Duluth to Sault Sainte Marie, have been traced and mapped by A. C. Lawson. To the south of Lake Superior the ancient shores have been systematically followed by F. B. Taylor. Many other observers have also contributed to this study, but not in such a methodical manner as those whose names have just been mentioned. Some of the problems that have presented themselves during this investigation have not yet been satisfactorily explained, but at least an outline of the Pleistocene history of the Laurentian basin may be presented with the understanding that it is to be modified as additional facts are obtained.

The most dramatic episode in the geological history of North America was the formation during Pleistocene time, of glaciers many hundreds of feet in thickness over the northern part of the continent. The ice advanced from the north and not only covered the Laurentian basin, but spread southward beyond the southern border of its watershed. The ice covered this region with various advances and retreats for thousands of years, and when it finally withdrew, the immediate ancestors of the present Great Lakes were born. There are several observations tending to the conclusion that during an interglacial time when the ice receded far north of its maximum limit, lakes were formed in the same basin, but in this connection there is little evidence to claim popular attention.

Previous to the Glacial epoch or the Great Ice age, as it is frequently termed, the region under review was an old land surface with rivers flowing across it to the sea. Its drainage system was well developed and the streams meandered through broad valleys, bounded in part by steep escarpments. In general relief, it must have resembled the upper portion of the Mississippi valley as it exists to-day, where the topography has not been modified by glacial action.

The conclusion that the Laurentian region was exposed to erosion for a long period previous to the Glacial epoch, is based on the character of the relief of the hard rock surface now covered in part by glacial deposits and on the fact that no sediments of younger date than the Carboniferous period, with the possible exceptions of terranes of Cretaceous age in portions of Minnesota, occur within its borders.

It may be suggested as a tentative hypothesis, that previous to the Glacial epoch the greater part of the Laurentian basin discharged its waters southward to the Mississippi, and that during the first advance of the ice from the north, the drainage was not obstructed so as to form important lakes. This suggestion rests in part on the fact that no lake deposits have yet been found beneath the lowest sheet of glacial débris lining the basin, — this negative evidence is of little weight, however, as such deposits, if they exist, would be mostly beneath the present lakes and therefore exceedingly difficult to discover, — and on the character of an ancient river valley leading south from the southern end of Lake Michigan, which is reported to be scored with glacial grooves, and obstructed by glacial deposits.[1] As will be noticed below, this same channel was also an outlet for the waters of the Lake Michigan basin in post-glacial times.

When the glaciers of the Glacial epoch were at their maximum, the drainage from the ice found a free escape southward, as is abundantly testified by immense deposits of gravel that were dropped by the overloaded glacial streams, as well as by numerous water-worn channels which are too large for the streams now occupying them and are without watersheds commensurate with their size.

As the ice sheet retreated, there came a time when its southern margin was north of the drainage divide, passing in an irregular east and west direction through Central New York and Central Ohio, and now parting the waters flowing south from those that find their way northward to the Laurentian lakes. When this occurred, lakes were formed between the margin of the ice and the high land to the south. These earlier lakes stood at various levels and discharged southward across the lowest depressions in their shores. Stream channels were excavated by the outflowing waters and became deeply filled with gravel and sand, but in many instances are still clearly traceable. One of these ancient channels starts near Fort Wayne, Indiana, leads southwest and afforded an escape for the waters that accumulated in the western portion of the Erie basin. A similar outlet at the south end of the Lake Michigan basin has already been referred to. Other points of discharge have been reported at other localities on the southern margin of the Laurentian basin.

As the ice occupying the Erie-Ontario basin withdrew northward, the lakes about its margin expanded and became united one with another. When the ice barrier between the two basins was broken the higher lake discharged into the lower one, and its former outlet leading south was abandoned.

When a single water body occupied the Erie-Ontario basin, the site of Niagara river was deeply submerged. When the water fell to the level of the Mohawk outlet, the two basins became divided and Niagara river was born. The river from the upper basin discharged across the lowest sag in its rim and cut back a deep gorge, until an old channel excavated in preglacial or possibly inter-glacial times, was discovered and the work of extending it renewed. When the falls shall have receded so as to drain

---

[1] Farther evidence seems to be needed, however, before the presence of a pre-glacial channel leading south from Lake Michigan, can be considered as definitely determined.

PLATE 18.

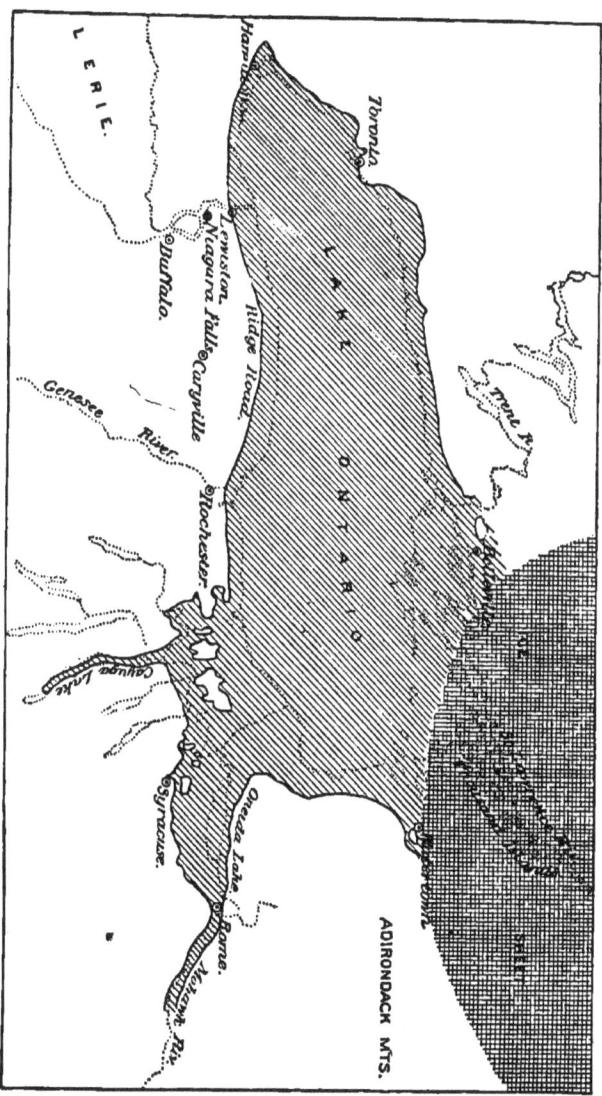

MAP OF LAKE IROQUOIS. (AFTER GILBERT.)

Modern hydrography in dotted lines. Ancient lake area shaded. Ice-sheet cross-shaded.

Lake Erie at a lower level than at present, the shore lines now forming about its margin will be abandoned and another line added to the records about its borders.

For a long period in the history of the Ontario basin, the outflowing water escaped through the Mohawk valley, New York, as has been shown by Gilbert, and the discharge of a large part of the Laurentian basin reached the sea by that channel. The series of well defined water-marks about the Ontario basin formed at this time, has been named the "Iroquois beach," by Spencer, and the ancient lake outlined by it is known as "Lake Iroquois." When the ice front retreated still farther northward, the present course of the St. Lawrence was uncovered, the Mohawk channel was abandoned, the water surface fell, and existing conditions were established.

-During various stages in the enlargement and subsequent contraction of the lakes about the southern margin of the Laurentide glacier, beaches were formed which in some instances, as has been shown by Frank Leverett, in Ohio, are continuations of the moraines deposited at the margin of the ice where lakes did not exist in front of it. In other instances moraines occur that are partially or wholly buried beneath lake sediments and mark the boundaries of the ice front where it was margined by water bodies.

At many localities where the former water markings are well preserved, they were made on low shores, and took the form of ridges resembling railroad embankments. The highest of these ridges marks the maximum limit of the water body about which it was formed. As the water fell the higher beaches were abandoned and others constructed at levels determined by lower outlets. When the borders of the lakes were of ice, shore records are wanting, but as stated above, buried moraines may mark the position of the dividing line between the water and the confining ice.

While the ancient beaches were in process of construction the abundant sediments carried into the lakes, were spread out as sheets of clay over the deeper portions of the basin, and at the same time the areas near shore received deposits of sand. Icebergs broke away from the glaciers forming the northern shores of the lakes, and floated over their surfaces, carrying stones which were dropped as the ice melted, and became imbedded in the clay on the bottom. These deposits surround the present Laurentian lakes and underlie them. About the borders of Lake Erie they appear as a stiff blue clay,—known to geologists as the "Erie clay," charged in some instances with large boulders of crystalline rock, — and as sheets of yellow sand, known as "delta sands," which rest on the clay, and are especially abundant where the mouths of ancient streams were located. About the shores of Lake Superior and frequently extending many miles inland, there are ancient clay deposits of a pink color, that were accumulated when the basin contained a much larger sheet of water than at present.

The beaches about the borders of the Laurentian lakes were originally horizontal, but as has been shown especially by Gilbert and Spencer, they are in many cases no longer in their original position. Changes in the elevation of the land have occurred and the beaches have been carried up or down with it.

The amount of change in level shown by the warping of the beaches about Lake Ontario is considerable, and illustrates the character of the slow upheavings and subsidences known to be in progress over wide areas of the earth's surface. It is stated by Gilbert[1] that "the old gravel spit near Toronto, belonging to what is known as the Davenport ridge, is forty feet higher than the contemporaneous gravel spit on which Lewiston is built; at Belleville, Ontario, the old shore is 200 feet higher than at Rochester; at Watertown, N. Y., 300 feet higher than at Syracuse; and the lowest point in Hamilton, Ontario, at the head of the lake, is 325 feet lower than the highest point near Watertown. From these and other measurements shown on Plate 18, we learn that the Ontario basin with its new attitude inclines more to the south and west than with the old attitudes." This general tilting has thrown the waters of Lake Ontario westward and flooded small tributary valleys so as to drown them and make miniature fiords.

Movements in the earth's crust were also in progress during the long period in which the ancient lakes of the Laurentian basin were making their various records, as is shown by the fact that the abandoned beaches do not all lie in planes parallel with each other.

The highest of the ancient beach lines about the north shore of Lake Superior, has an elevation of about 600 feet above the present lake, as has been determined by A. C. Lawson.[2] The beaches at lower levels are

---

[1] "The history of Niagara river," in Sixth Annual Report of the Commissioners of the State Reservation at Niagara, Albany, N. Y., 1890, p. 69. Reprinted in Ann. Rep. Smithsonian Institution, 1890, pp. 231–257.

[2] "Sketch of the Coastal Topography of the North Side of Lake Superior," in 20th Ann. Rep., Minnesota, Geol. and Nat. Hist. Surv., pp. 181–289.

approximately parallel with it. Observations on the amount of deformation that this beach has suffered, are not as extended as could be desired,
but near its western extension there is evidence of a change of level of
about one foot per mile.

Recent observations by F. B. Taylor[1] in the region adjacent to Lake
Superior on the south, have shown that ancient beaches may be clearly
recognized at many places between Duluth and Sault Sainte Marie. The
facts recorded by Taylor supplement in a very interesting manner the
work of Lawson on the northern side of the same basin, although farther
study is necessary before the entire history of the great predecessor of Lake
Superior can be written. At the south, the highest beach has an elevation of from 512 to 588 feet above Lake Superior, or from 1014 to 1190
feet above the sea.

Taylor suggests that when the entire outline of the highest beach at
the north shall have been traced, it will be found that there were straits
connecting the Superior basin with that of Hudson Bay. This would
imply a submergence of a very large portion of the North American continent to a depth of over a thousand feet.

The erosion produced by the movement of ice sheets many hundreds
of feet thick, over the Laurentian basin, modified and subdued the previous relief, and the débris left when the ice melted covered the country
with a sheet of superficial deposits to such a depth that the character
of the underlying hard-rock topography is only occasionally revealed.
The depth of these glacial deposits over great areas, as in Michigan
and Wisconsin, is from one to two hundred feet, but is probably of less
average thickness in Ohio and New York. All pre-glacial drainage channels
were either obstructed or obliterated and a new surface given to the land.
The drainage was thus rejuvenated and is still immature. The effects of
glacial plantation and of glacial deposition, in forming the basins of the
present Laurentian lakes, has been pointed out in discussing the origin of
lake basins.

In this brief sketch I have endeavored to show that the history of
the Laurentian basin includes a study of the hard-rock topography as it
existed previous to the Glacial epoch ; the disturbances and changes in
drainage produced by the ice invasion and by movements of elevation
and depression; the obstruction of the ancient waterways by glacial
deposits ; and the origin of new channels of discharge, as the glaciers

---

[1] " A reconnoissance of the abandoned shore lines of the south coast of Lake Superior,"
in Am. Geol., Vol. 13, 1894, pp. 365–383. See also more recent papers in the same journal.

passed away, — all of these links in the complex history have not been completely worked out, and this attractive field is still open to the geologist and geographer.

In conclusion, it is but fair to state that while the history of the Laurentian basin outlined above will, I believe, be accepted as in the main correct by most geologists of the United States, whose attention has been directed to the subject, it is widely at variance with the conclusions of at least two Canadian geologists. Sir J. William Dawson maintains, if I understand his hypothesis correctly, that the sea, laden with icebergs, invaded the Laurentian basin in Pleistocene times, and that the moraines and other deposits occurring in it and over a wide extent of adjacent country, and believed by most observers to be of glacial origin, are shore accumulations, and that icebergs and floe-ice played an important part in their formation.

The ancient beaches about the Laurentian lakes, while considered as true shore lines by Spencer, are thought by him to have been formed at sea-level during a time of continental submergence, and that the ocean had free access to the basin.

It may be that in these summary statements I do injustice to the views of the gentlemen referred to, but the conclusions indicated are so widely at variance with a vast body of consistent evidence gathered by a score or more of skilled observers, and is so directly opposed to my own observations, both of living glaciers and of the records of past glaciation, that they do not seem at present to be open to profitable discussion.

A subsidence of the eastern border of the continent during the later stages of the Glacial epoch, or following its close, throughout a belt widening from New York city northward, and including the valley of Lake Champlain, is well known. When the studies leading to this conclusion are extended to the basins of the Laurentian lakes, however, not only is there an absence of salt-water shells and other evidences of marine occupation, but, seemingly, positive evidence of lacustral condition.

The region to the north of Lake Superior has not been sufficiently studied to admit of an opinion being reached in reference to the questions just considered, from the records there obtained. It may be found that the highest shore-line in the Superior basin was formed by a water body in direct communication with the sea to the north, as suggested by Taylor. Should this hypothesis be sustained, it would add an interesting chapter to the history of the Superior basin, and render a review desirable of the evidence of a similar nature in the eastern portion of the region now drained by the St. Lawrence.

The views of Dawson and Spencer are set forth in the publications mentioned in the following footnote,[1] and should be attentively studied by all who undertake to read the history of the Laurentian basins from the original records in order that their conclusions may be fairly tested.

LAKE AGASSIZ.

At the time the remarkable changes described above were taking place in the Laurentian basin, there were corresponding revolutions in the geography of the region to the northwest which now drains to Lake Winnepeg and thence through Nelson river to Hudson bay.

It will be readily seen on glancing at a map of Canada, that if a glacier of the continental type should advance southward from the Hudson bay region, the drainage would be obstructed and a lake formed over the country of mild relief surrounding Lake Winnepeg and the Lake of the Woods, and extending southward through the Red River valley, far into Minnesota. Such a lake would discharge southward, and contribute its surplus waters to the Mississippi. Should the hypothetical glacier referred to advance until it occupied all of the Winnepeg basin, the lake about its southern margin would be obliterated, and there would be free drainage to the Gulf of Mexico. Should the glacier then retreat to the north of the divide now separating the waters flowing southward to the Gulf of Mexico from those flowing northward to Hudson bay, a lake would be born about the margin of the ice, and would increase northward as the ice retreated. When a channel leading northward was uncovered and rendered available as an outlet for the lake, the ponded waters would have their level lowered and their area contracted.

The study of the Pleistocene records in the Red River valley and thence northward in Manitoba, has shown that changes very similar to those postulated above actually occurred.

The evidence of the former existence of a large lake in the Red River valley was observed as far back as 1823 by Keating, the geologist of the first scientific expedition to that region. Subsequent contributions to this investigation have been made by several observers, and notably by

---

[1] J. W. Dawson, "The Canadian Ice Age," Montreal, 1893 ; J. W. Spencer, "The Deformation of Iroquois Beach and Birth of Lake Ontario," in Am. Jour. Sci., ser. 3, vol. 40, 1890, pp. 443–451 ; J. W. Spencer, "Deformation of the Algonquin Beach and the Birth of Lake Huron," in Am. Jour. Sci., ser. 3, vol. 41, 1891, pp. 12–21 ; J. W. Spencer, "Post-Pleistocene Subsidence versus Glacial Dams," in Geol. Soc. Am. Bull., vol. 2, 1891, pp. 465–474.

Gen. G. K. Warren, who first explained the origin of the valley now occupied by Lake Traverse, Big Stone lake, and the Minnesota river, by showing that it was excavated by a stream flowing to the Mississippi from a former lake to the north. This ancient river, whose source has long since been sapped by northward drainage, has been named River Warren, after its discoverer.

The great lake that formerly flooded the Winnepeg basin, and during its highest stage overflowed through River Warren, has been named Lake Agassiz, by Warren Upham, in honor of Louis Agassiz. Practically all of the facts and conclusions here presented concerning the history of that remarkable lake, have been made known through the long-continued and skillful investigations of Upham, under the auspices, at different times, of the geological surveys of Minnesota, the United States, and Canada,[1] respectively.

The Red River of the North rises in the western part of Minnesota, and receives the tribute of Lake Traverse, situated on the Minnesota-Dakota boundary, and at the southern limit of the country formerly flooded by Lake Agassiz. From Lake Traverse the present drainage is northward through narrow channels sunken in the sediments of the former lake. Between the streams there are broad, nearly level, inter-stream spaces, forming typical examples of new-land areas, on which shallow ponds form during rainy seasons. About the borders of this broad, level extent of prairie land, now transformed into wheat fields, there are gravel ridges which mark the surface level of the former lake at various stages. These ancient beaches have been traced northward and found to diverge toward the northeast and northwest when the central area of the old lake was approached, and have been mapped so as to show approximately the extent of the water body that built them. By patiently following these ancient shore-lines, it has been demonstrated that Lake Agassiz covered a region about 110,000 square miles in area. Its diameter from north to south was 675 miles, and from east to west, in the wider portions, varied from 225 to 300 miles. It was the largest of the Pleistocene lakes of North America thus far discovered, and exceeded the combined areas of the present Laurentian lakes. The rim of its hydro-graphic basin embraced a region not less than half a million square miles in area. At the site of Lake Winnepeg the ancient lake was 600 feet deep.

---

[1] A report on these investigations appeared in the Geol. and Nat. Hist. Survey of Canada, Ann. Rep., vol. 4, 1888-9, pp. 1-150 E, and a monograph on the same subject is soon to be issued by the U. S. Geol. Survey.

One of the most interesting discoveries in connection with the beaches of Lake Agassiz, is that they are no longer horizontal, and besides do not lie in plains that are parallel one with another. The highest water line when followed northward has been found to rise at the rate of 200 feet in 300 miles. There are five beaches that are especially prominent and mark a lingering of the lake surface at their respective horizons. The highest of the series, known as the Herman beach, when traced northward from the southern end of the Red River valley, has been found to divide into several beaches at different levels; the vertical intervals between the division increasing northward. The meaning of this fact seems to be that the land was rising at the north at the time the beaches were formed and at the same time the surface of the lake was lowered by reason of the opening of new outlets.

To the north of Lake Winnepeg the higher of the ancient beaches are absent and the lower ones difficult to trace. The country still farther toward Hudson bay is low and does not present a barrier that under any plausible hypothesis could have been made to act as a dam to retain the waters of Lake Agassiz. What then could for a time have reversed the drainage and led to the formation of a lake over a hundred thousand square miles in area?

The origin of Lake 'Agassiz as explained by Upham, is in harmony with the history of the former lakes of the Laurentian basin. It is supposed to have owed its origin to the presence of a vast ice sheet over the Hudson bay region which dammed the northward drainage of the Winnepeg basin and caused the waters to rise until an outlet was found at the south and River Warren began to flow. When the ice retreated, new outlets at lower levels became available at the north and the waters fell, but lingered for a time at the horizon of each of the various beaches that have been referred to, at lower levels than the Herman beach.

There are facts in connection with the ancient floods of the Laurentian and Winnepeg basins, which seem to indicate that the weight of the ice during the Glacial epoch caused the land to subside, and that when the ice melted an upward movement was initiated. These movements, and also the attraction of the ice body to the north of Lake Agassiz, have been thought to explain the gradual rise of the beaches when traced northward.

The strange transformation that the Winnepeg basin underwent in Pleistocene times, leads one to wonder if in the region now drained by Mackenzie river, and occupied in part by Great Slave and Great Bear lakes, there may not be equally wonderful records awaiting the coming of the patient inquirer.

PLEISTOCENE LAKES OF THE GREAT BASIN.

During the time of great climatic changes that witnessed the birth, growth, and decadence of the great lakes of the Laurentian and Winnepeg basins, described above, equally important fluctuations occurred in the lakes of the Arid region. Many of the valleys of Utah and Nevada, and of adjacent areas both north and south, that are now parched and desert-like throughout the year, were then flooded, and in some instances filled to the brim so as to overflow. All of the enclosed lakes west of the Rocky mountains were then of greater size than at present and underwent marked changes in sympathy with the advance and retreat of glaciers on neighboring mountains, and had their oscillations controlled by the same causes, viz., variations in precipitation, evaporation, and temperature.

Of these numerous water bodies there were two of broad extent which may be taken as types of their class and will serve to give an epitome of the history of their time. The two ancient lakes referred to are Bonneville and Lahontan [1] and are represented on the map forming Plate 19.

Lake Bonneville was named by Gilbert in honor of Captain B. L. E. Bonneville, U.S.A., who made a bold exploration into the wilds of the Rocky mountains in 1833, and was the first person to gather reliable information concerning the region formerly occupied by the great lake now bearing his name. The reader will perhaps have an additional interest in the following sketch, when he recalls the "Adventures of Captain Bonneville," so graphically described by Washington Irving.

Lake Lahontan first received definite recognition in the reports of the 40th Parallel survey under the direction of Clarence King, and was named after Baron LaHontan, one of the early explorers of the Mississippi valley. Why LaHontan's name should have been thus connected with a region more than a thousand miles beyond his farthest camp, in preference to the names of men who boldly crossed and recrossed the land referred to when it was a trackless desert infested with roving bands of savages, I must leave to others to explain.

As shown on the accompanying map, Plate 19, Lake Bonneville occupied the basin in which Great Salt lake now lies, on the east side of the

---

[1] Clarence King, U. S. Geol. Exploration of the 40th Parallel. Vol. 1, 1878, pp. 490–529. — G. K. Gilbert, "Lake Bonneville." U. S. Geol. Surv., Monograph No. 1, 1890. — I. C. Russell, "Lake Lahontan." U. S. Geol. Surv., Monograph No. 11, 1885.

PLEISTOCENE LAKES OF THE GREAT BASIN.

This map is incomplete, as the entire area has not been studied.

Great Basin, while Lake Lahontan flooded a series of irregular valleys on the west side of the same great area of interior drainage and is now represented by Pyramid, Winnemucca, Walker, Carson, and Humboldt lakes, Nevada, and by Honey lake, California.

These two ancient lakes were contemporaries, and, although differing in their histories, bear similar testimony in reference to climatic changes and supplement each other's records in a remarkable manner. Their hydrographic basins joined each other in north-eastern Nevada, for a distance of about twenty-five miles, and together occupied the entire width of the Great Basin. Lake Bonneville received its water supply from the Wasatch and Uinta mountains, then snow-clad throughout the year and holding glaciers of the Alpine type in many of their valleys. Several of the ice streams on the precipitous western slope of the Wasatch mountains reached nearly to the ancient lake which washed the base of the range, and one of them was prolonged for a short distance into its waters. Lake Lahontan derived its principal water supply from the Sierra Nevada, which formed the western rim of its drainage basin for a distance of 250 miles, and, like the eastern borders of the Bonneville basin, was glacier-covered.

Lake Bonneville at the time of its maximum extension had an area of 19,750 square miles, and a hydrographic basin 52,000 square miles in area. The more irregular water surface of Lake Lahontan was 8,422 square miles in area, and occupied the lowest depressions in a hydrographic basin containing 40,775 square miles. The great size of the hydrographic basins of these lakes in comparison with their extent of water surface, is a noteworthy feature. The ratio of the extent of lake surface to area of hydrographic basin in the case of Lake Bonneville was as 1 to 2.6, and in the case of Lake Lahontan about 1 to 5. The corresponding ratios in the basin of Lake Superior are as 1 to 1.72; and for the combined Laurentian lakes as 1 to 3.19. The small extent of the ancient lakes of the Great Basin in comparison with the areas draining to them, more especially in the case of Lake Lahontan, indicates that the climate of their time was not markedly humid.

The maximum depth of Lake Bonneville as recorded by beach lines on the mountain forming its shores, and on the precipitous islands now rising in Great Salt lake, was 1050 feet. The greatest depth of Lake Lahontan was 886 feet.

The most striking difference in connection with these two ancient seas is in reference to overflow. The waters of Lake Bonneville rose

until they found an outlet and escaped through a channel leading northward from Cache valley, in Utah and Idaho, to Snake river and thence to
the Columbia. The outflowing stream at its source crossed incoherent
alluvial deposits and rapidly cut down a channel of discharge to a depth
of 370 feet, thus lowering the lake by that amount. During this episode
in its history the lake was fresh, but at later stages, when its surface fell
below the level of the bottom of the channel of discharge, it became
saline. The water supply of Lake Lahontan was less abundant and it
never rose so as to find an outlet. Its waters were perhaps brackish
during its higher stages, and became saline and alkaline as concentration
progressed.

Each of these lakes had two high-water stages, separated by a time of
low water and probably of complete desiccation. The second high-water
stage in each instance was the more marked of the two. These fluctuations are indicated in the following diagram of the rise and fall of Lake
Lahontan.

Each lake spread out two sheets of fine, evenly-laminated clays, sepa-

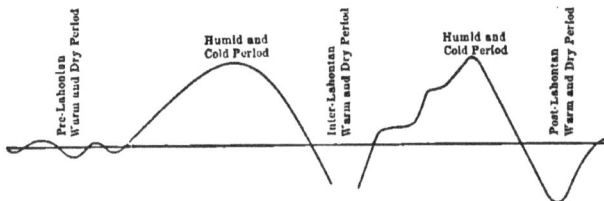

FIG. 8. — DIAGRAM SHOWING THE RISE AND FALL OF LAKE LAHONTAN.

rated, at least about their borders, by deposits of coarse gravel and sand
washed in from the adjacent slopes during the inter-lacustral time of low
water.

There are many reasons for concluding that the two high-water stages
recorded by beach lines and by sedimentary deposits in the basins of lakes
Bonneville and Lahontan, correspond in time with two of the periods of
glaciation recorded in the Laurentian basin. Two periods of marked
advance separated by a time of retreat, are also indicated by the glacial
records in the cañons of the Sierra Nevada.

The waters of both Bonneville and Lahontan underwent many minor
fluctuations of level as is the rule with all enclosed lakes. The terraces,
embankments, deltas, etc., constructed about the shores of Lake Bonneville are on a grander scale than in the basin of its companion lake, for

the reason that it was the larger of the two water bodies and had a more regular outline, thus giving the wind a better opportunity to act on its waters, and also because it was held at a definite level for a long period, or rose to the same horizon at various times, on account of its having an outlet.

The highest water line about Lake Bonneville, named the "Bonneville beach," is conspicuous not so much on account of its strength as for the reason that it marks the dividing line between rain sculpture on the higher portions of the bordering mountains and the characteristic topography due to the work of waves and currents on their lower slopes. The channel of discharge was lowered until a sill of resistant limestone was reached which determined the horizon of the strongest and best developed terraces and embankments in the basin. A well defined beach at this horizon is known as the "Provo beach," the name being derived from the town of Provo, Utah, which stands on a broad delta formed by the sediment of Provo river, when the lake stood at the horizon of its lowest point of discharge. The wave-built structures marking the Provo stage are on a magnificent scale and are still almost as fresh in appearance and perfect in form as if abandoned by the waves but yesterday. In the Lahontan basin the shore topography was never strongly pronounced. Fluctuations of level were not controlled by an outlet, and the numerous islands and headlands diminished the influence of the wind and checked the action of waves and currents.

The chemical histories of lakes Bonneville and Lahontan are fully as instructive and of as great interest as their physical changes. In this connection, the basin of Lake Lahontan has been found to exceed its companion in the completeness of its records. The escape of the waters of Lake Bonneville insured its freshness during a part of its history. The absence of an outlet for the waters of Lake Lahontan led to a high degree of concentration.

When lake waters are concentrated by evaporation the first substance to be precipitated, as previously described, is calcium carbonate. About the shores of Lake Bonneville there are in favorable localities, considerable deposits of this substance in the form of coral-like incrustations known as calcareous tufa. It appears on rocky points and forms a cement for gravel and sand on the outer borders of some of the terraces, but is insignificant in amount and simple in character, when compared with the truly immense accumulations of a similar nature in the Lahontan basin.

The precipitation of calcium carbonate from lake waters takes place principally in two ways; it may separate in the open lake and fall to the bottom in a finely divided state and become mingled with mechanical sediments so as to form marls, or it may be precipitated where solid rocks occur and cover them with a dense incrustation. The ability of ordinary surface waters to dissolve calcium carbonate, depends mainly on the carbonic acid gas they hold in solution. Lake waters lose their dissolved gases most rapidly where they form breakers along the shore, as in such instances they are most thoroughly aërated. For this reason, the boldest headlands are apt to receive the heaviest deposits of tufa when the waters dashed against them became concentrated. It is at such localities that the principal deposits of tufa in the Bonneville basin occur. It happens also that calcium carbonate has a tendency to accumulate about solid bodies, not only because they afford a stable support, but. for the additional reason that points and angles induce crystallization. Calcareous tufa was deposited in vast quantities about the shores of Lake Lahontan wherever there were rocky slopes and in increasing abundance from an horizon high up on its borders down to the deepest point now exposed. The fluctuations of level in Lake Bonneville were recorded principally by beaches and embankments of mechanical origin ; similar changes in Lake Lahontan are made known by tufa deposits of chemical origin.

The tufa of the Lahontan basin presents three main varieties, each of which is composed of concentric layers as is shown in Plate 21. The smaller divisions seem to indicate minor changes in the chemistry, and perhaps also fluctuations in the temperature, of the water from which they were precipitated. The three principal varieties have been named in the order of their formation, Lithoid, Thinolitic, and Dendritic tufa. Lithoid tufa is a compact stony substance with a granular texture ; Dendritic tufa has an open structure and resembles a mass of branching twigs turned to stone ; and Thinolitic tufa, shown in Plate 22, is composed of well defined crystals to which the name Thinolite was given by Clarence King. The composition of each of these varieties is the same. They are composed of calcium carbonate with usually some slight amount of impurities. Their wide variation in structure and general appearance, is due to differences in the condition of the lake waters at the time of their formation.

About Pyramid lake, where the Lahontan tufas are usually well displayed, the first or Lithoid variety reaches a height of 500 feet, the Thinolitic 110 feet, and the third or Dendritic variety, 320 feet above the

TUFA TOWERS ON THE SHORE OF PYRAMID LAKE, NEVADA.

surface of the present lake. The relation of the tufa deposits and the terraces with which they are associated, are shown in the following diagram.

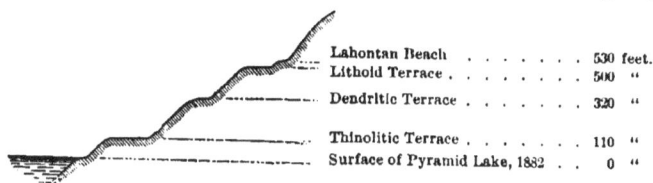

Lahontan Beach . . . . 530 feet.
Lithoid Terrace . . . . 500 "
Dendritic Terrace . . . . 320 "

Thinolitic Terrace . . . . 110 "
Surface of Pyramid Lake, 1882 . . 0 "

FIG. 9. — DIAGRAM SHOWING THE RELATION OF THE TERRACES OF LAKE LAHONTAN
TO PYRAMID LAKE.

The Lithoid tufa near its upper limit is seldom over eight or ten inches thick, but increases to ten or twelve feet on the lower slopes. The Thinolite is usually from six to twelve feet thick. The Dendritic variety is the heaviest of all and frequently appears on steep slopes in imbricated layers from fifty to sixty feet thick. In some favorable locality the entire tufa deposits have a thickness of at least eighty feet, and in rare places near the surface of Pyramid lake and partially concealed by its waters, there is evidence that these deposits are still more massive. The total amount of calcium carbonate deposited from the ancient lake can only be estimated in millions, if not billions of tons.

Every island and rocky crag that rose in Lake Lahontan became a center of accumulation for tufa deposits and was transformed into strange and frequently fantastic shapes by the material precipitated upon it. Now that the waters of the ancient sea have disappeared, these structures stand in the desert valleys like the crumbling ruins of towers, castles, domes, and various other shapes, in keeping with the desolation surrounding them. The finest examples of these water-built structures, some of them a hundred feet or more in height, occur about the border of Pyramid and Winnemucca lakes (Plate 20), or rising from their bottoms and still wholly or in part submerged. The islands in Pyramid lake are sheathed from base to summit with these deposits and their precipitous sides given a convex outline, owing especially to the vast deposits of Dendritic tufa, which was precipitated most abundantly midway up the slopes. The most remarkable of these islands, and the one from which the lake derives its name, is shown in the sketch forming Plate 23. When the tufa towers and castle-like piles are broken, the concentric layers of which they are composed are revealed and fill one with wonder at the vast amount of material they contain, as well as attract the eye on account of the delicacy and beauty of their structure. Nowhere else in this country, and so far as reported, nowhere else in the world, are rocks formed of precipitates from lake waters so magnificently displayed as in the desert valleys of Nevada.

The fascination of the weird and frequently wonderfully impressive scenery of the region formerly submerged beneath the waters of Lake Lahontan, is enhanced, at least to the geologist, by the fact that there is yet an unsolved mystery connected with the tufa deposits that start out as strange, gigantic forms from the desert haze, as one slowly traverses those bitter, alkaline lands.

It is believed that we understand how the more compact and stone-like variety of tufa was deposited, since similar accumulations are formed where waters saturated with calcium carbonate deposit that salt on account of the loss of carbonic acid. The Dendritic tufa may also have been precipitated in a similar manner, or perhaps through the agency of low forms of plant life. The mode of origin of the tufa with well-defined crystals, however, is still unknown, although both geologists and chemists have sought diligently to discover the secret of its formation. The open cellular structure of the crystals, as well as their forms, suggest that they are pseudomorphs, that is, having a false form, or a form not assumed by calcium carbonate on crystallizing, but resulting from the alteration or replacement of some other mineral. This suggestion only removes the difficulty one step farther, however, since the nature of the original mineral is still unknown. A more definite statement of this problem may be found in a special report on Thinolite, by E. S. Dana, who has put the matter in a clearer light than had previously been done.[1]

One of the most remarkable facts in connection with the history of the Lahontan basin, is that the present lakes within it, which might be supposed to be remnants of the ancient water-body left by incomplete evaporation, and therefore intensely saline, are in reality scarcely more than brackish. As shown in the table of analyses of saline lakes given on page 72, Pyramid, Winnemucca, and Walker lakes, the representative water bodies now existing in the Lahontan basin, carry only a small fraction of one per cent of saline matter in solution. We know that Lake Lahontan did not overflow. All of the saline matter carried into it, therefore, must still be retained in its basin. The vast quantity of various salts, and especially of sodium chloride, sodium sulphate, and sodium

---

[1] "Crystallographic Study of the Thinolite of Lake Lahontan," Bulletin No. 12, U. S. Geol. Survey.

PLATE 21.

TUFA CRAG AT ALLEN'S SPRINGS, NEVADA, SHOWING SUCCESSIVE DEPOSITS.

carbonate thus concentrated, is indicated by the weight of the calcareous tufa lining the basin. In ordinary river waters, as already shown, the calcium carbonate is about the same as the amount of all other salts in solution. It follows, therefore, that the more soluble salts contributed to Lake Lahontan must have been equal in weight to the tufa deposits just described. Such a vast quantity of saline matter, if contained in the present lakes, would make them concentrated brines. The question is, what has become of the more soluble salts contributed to the waters of the ancient sea?

A lake may occasionally evaporate to dryness, or exist as a playa lake for a long period, that is, expanding during rainy seasons and becoming desiccated either during dry seasons, or occasionally in years of unusual aridity. Under such conditions its contained salts would be precipitated and become buried or absorbed by mechanical sediments, so that when a change of climate permitted the existence of a perennial lake in the same basin, it would be fresh, or essentially so. This is what seems to have occurred in the Lahontan basin. The old lake was probably evaporated to dryness and the precipitated salts buried beneath playa clays, and when a change to slightly more humid conditions permitted of the birth of the present lakes, a new cycle was begun.

From analyses of the waters flowing into the present lake of the Lahontan basin, it has been estimated that under existing conditions they would acquire their present degree of salinity in about 300 years. It seems to follow from this study that during a long term of years, ending about 300 years ago, the climate of Nevada was so intensely arid that no perennial lakes could exist within her borders.

An account of the physical and chemical histories of the ancient lakes of Utah and Nevada should be followed by a description of the plants and animals that found a home on their shores, but unfortunately our information in this connection is vague.

The sediments of lakes Bonneville and Lahontan, unlike many other lake-beds, are extremely poor in vegetable fossils. As the conditions for the preservation of such remains were favorable, and as an extended search has failed to unearth so much as a single leaf or a single water-logged tree-trunk from their sediments, it may reasonably be concluded that their shores were not forested, and were probably even more barren and desolate than at the present day. This result cannot be considered as surprising in view of the great fluctuation of climate that the Great Basin experienced in Pleistocene times.

Of the remains of vertebrates, the bones of the mastodon or mammoth, and of the ox, camel, and horse have been found in the sediments of Lake Lahontan, together with a single undetermined fish. The bones of a musk-ox were obtained near Salt Lake City under such conditions that it is believed they were buried in the upper strata of the Bonneville sediments. The basins of contemporaneous lakes in Oregon, have yielded vertebrate fossils more abundantly, but concerning these there are differences of opinion as to their age. It is probable that some of them at least, and perhaps the larger portion, were washed out of older deposits and accumulated in the basin where they are now found.

In the sediments of both Bonneville and Lahontan there are many species of fresh-water shells, but these are usually small individuals, and appear to have lived under uncongenial conditions.

The remains of animal life do not seem to point to any very definite conclusion. We are led to believe from all of the evidence available, however, that the climate of the lake period was cold and changeable, and consequently uncongenial to either plant or animal life. The interlacustral epoch was probably a time of high temperature and aridity. The large animals whose bones have been discovered may have been forced to migrate owing to wide-reaching climatic changes, and were perhaps only temporary visitors to the region where they succumbed to adverse conditions.

The mastodon and mammoth roamed over nearly the whole of North America during Pleistocene times, but have since become extinct. The camel is no longer found on this continent, and the horse was extinct before the coming of the white man. The musk-ox is now found only far to the north. The extinction of some of these large animals, and the scattering of others to distant regions, suggests the lapse of a long period of time since they lived together where their remains are now found, and also points to great changes in climatic and other elements of their environment.

Of the presence of man on the shores of lakes Bonneville and Lahonton the records are silent.

LAKES OF THE REMOTE PAST.

The presence of the bones of large animals in the sediments of lakes Bonneville and Lahontan naturally leads one to look farther back in the earth's history, to the deposits of other lakes from which a vast menagerie of strange and frequently gigantic forms have been made known by the labors of American paleontologists.

Immediately preceding the "Great Geological Winter," as the Glacial epoch has been termed, when half of the North American continent was sheathed in ice, there was a period of genial climate when vegetation, as varied and beautiful as that of the Mississippi valley to-day, extended far north and reached the vicinity of the pole itself. During different epochs in this geological summer, known as the Tertiary period, vast fresh-water lakes existed in the Cordilleran region, several of which were far more extensive than any lakes now known. In some of these vast inland seas several thousand feet of sediments were laid down. In these deposits we find in abundance the impressions of leaves that were blown from the land, or washed in by tributary streams, and the bones of many large mammals, whose homes were along the lake shores and on neighboring forest-covered hills.

All trace of the shore topography of the Tertiary lakes has disappeared, and in many instances the beds of sand, clay, and volcanic dust deposited over their bottoms have been upheaved into mountain ranges, and deeply dissected by erosion. Their histories can only be deciphered from the records in their sediments. Their story deals largely with the structure, habits, and development of vertebrate animals, and must be left to those skilled in that branch of study.

Beyond the Tertiary period, and so remote from our own time that humble forms of mammalian life had only just appeared on the earth, were the Jurassic and Triassic periods. In this Mesozoic time, or middle age of the earth, lakes also existed, and in their sediments the skeletons of another striking and grotesque assemblage of strange forms were preserved. The magic wand of modern science has brought forth from these long-silent tombs a wonderful procession of gigantic reptiles, the like of which has not since existed on the earth.

Still more remote were the lakes and swamps of the Carboniferous period. The oldest records of air-breathing vertebrates yet discovered are the bones of reptiles found by Dawson in hollow-tree trunks that stood in the fresh-water swamps of Nova Scotia during the time our continent was green with the ferns and club-mosses of the Coal period. With these bones are mingled the shells of land-snails, the earliest of their class yet found.

In deposits of cannel coal formed in fresh-water ponds in the great coal swamps of Ohio, Newbery discovered a large number of species of fishes and amphibians, in a beautiful state of preservation.

Farther back still in the records of the past are other fragments of the earth's history sealed up and preserved in lake deposits. The heavy beds of sandstone composing the Catskill mountains, and forming a part of the Devonian system, contain shells which resemble the covering of fresh-water mollusks, and may indicate that the sands in which they were buried are of lacustral origin. Here the evidence of terrestrial lakes seems to end. What inland water bodies existed in remote Silurian, Cambrian, and Algonkian times, remains to be discovered.

A CHARACTERISTIC SPECIMEN OF THINOLITE.

SUPPLEMENT.

The advance made in the study and in the interpretation of the meaning of topographic forms, has been so great, especially in America, during the present decade, that I am sure the reader will be interested in the writings of those who have made this important departure from old methods. The recognition that lakes are transient features of the ever-changing earth's surface and come and go during cycles of topographic development, was first clearly set forth in a brief paper by W. M. Davis,[1] which is here reproduced.

THE CLASSIFICATION OF LAKES.

Several years ago I presented to the Boston Society of Natural History a paper on the classification of lake-basins, in which the many varieties of lakes were grouped under three heads, according as they were made by constructive, destructive, or obstructive processes. The first heading included lakes made by mountain-folding and other displacements; the second consisted chiefly of basins of glacial erosion; the third contained the greatest number of varieties, such as lakes held by lava, ice, and drift barriers, delta and ox-bow lakes, and some others. The classification proved satisfactory, in so far as it suggested a systematic arrangement of all kinds of lakes that have been described; but it now appears unsatisfactory, inasmuch as its arrangement is artificial, without reference to the natural relations of lakes to the development of the drainage systems of which they are a part. A more natural classification is here presented in outline.

When a new land rises from below the sea, or when an old land is seized by active mountain-growth, new rivers establish themselves upon the surface in accordance with the slopes presented, and at once set to work at their long task of carrying away all of the mass that stands above sea-level. At first, before the water-ways are well cut, the drainage is commonly imperfect: lakes stand in the undrained depressions. Such lakes are the manifest signs of immaturity in the life of their drainage system. We see examples of them on new land in southern Florida; and on a region lately and actively disturbed in southern Oregon, among the blocks of faulted country described by Russell. But as time passes, the streams fill up the basins and cut down the barriers, and the lakes disappear. A mature river of uninterrupted develop-

[1] Science, vol. 10, 1887, pp. 142, 143.

ment has no such immature features remaining. The life of most rivers is, however, so long, that few, if any, complete their original tasks undisturbed. Later mountain-growth may repeatedly obstruct their flow ; lakes appear again, and the river is rejuvenated. Lake Lucerne is thus, as Heim has shown, a sign of local rejuvenation in the generally mature Reuss. The head waters of the Missouri have lately advanced from such rejuvenation; visitors to the National Park may see that the Yellowstone has just regained its former steady flow by cutting down a gate through the mountains above Livingston, and so draining the lake that not long ago stood for a time in Paradise valley. The absence of lakes in the Alleghany mountains, that was a matter of surprise to Lyell, does not indicate any peculiarity in the growth of the mountains, but only that they and their drainage system are very old.

The disappearance of original and mountain-made lakes is therefore a sign of advancing development in a river. Conversely, the formation of small shallow lakes of quite another character marks adolescence and middle life. During adolescence, when the head-water streams are increasing in number and size, and making rapid conquest of land-waste, the lower trunk-stream may be overloaded with silt, and build up its flood-plain so fast that its smaller tributaries cannot keep pace with it : so the lakes are formed on either side of the Red River of Louisiana, arranged like leaves on a stem; the lower Danube seems to present a similar case. The flood-plains of well-matured streams have so gentle a slope that their channels meander through great curves. When a meander is abandoned for a cut-off, it remains for a time as a crescentic lake. When rivers get on so far as to form large deltas, lakes often collect in the areas of less sedimentation between the divaricating channels. Deltas that are built on land where the descent of a stream is suddenly lessened and its enclosing valley-slopes disappear, do not often hold lakes on their own surface ; for their slope is, although gentle, rather too steep for that : but they commonly enough form a lake by obstructing the stream in whose valley they are built. Tulare Lake in southern California has been explained by Whitney in this way.

The contest for drainage area that goes on between streams heading on the opposite slopes of a divide sometimes produces little lakes. The victorious stream forces the divide to migrate slowly away from its steeper slope, and the stream that is thus robbed of its head waters may have its diminished volume clogged by the fan-deltas of side-branches farther down its valley. Heim has explained the lakes of the Engadine in this way. The Maira has, like an Italian brigand, plundered the Inn of two or more of its upper streams and the Inn is consequently ponded back at San Moritz and Silvaplana. On the other hand, the victorious stream may by this sort of conquest so greatly enlarge its volume, and thereby so quickly cut down its upper valley, that its lower course will be flooded with gravel and sand, and its weaker side-streams ponded back. No cases of this kind are described, to my knowledge, but they will very likely be found; or at least we may expect them to appear when the northern branches of the Indus cut their way backwards through the innermost range of the Himalaya, and gain possession of the drainage of the plateaus beyond; for then, as the high-level waters find a steep outlet to a low-level discharge, they will carve out cañons the like of which even Dutton has not seen, and the heavy wash of waste will shut in lakes in lateral ravines at many points along the lower valleys.

In its old age, a river settles down to a quiet, easy, steady-going existence. It has overcome the difficulties of its youth, it has corrected the defects that arose from a period of too rapid growth, it has adjusted the contentions along the boundary-lines of its several members, and has established peaceful relations with its neighbors : its lakes disappear, and it flows along channels that meet no ascending slope on their way to the sea.

Certain accidents to which rivers are subject are responsible for many lakes. Accidents of the hot kind, as they may be called for elementary distinction, are seen in lava-flows, which build great dams across valleys : the marshes around the edge of the Snake river lava-sheets seem to be lakes of this sort, verging on extinction : crater lakes are associated with other forms of eruption. Accidents of the cold kind are the glacial invasions: we are perhaps disposed to overrate the general importance of these in the long history of the world, because the last one was so recent, and has left its numerous traces so near the centers of our civilization ; but the temporary importance of the last glacial accident in explaining our home geography and our human history can hardly be exaggerated. During the presence of the ice, especially during its retreat, short-lived lakes were common about its margin. We owe many prairies to such lakes. The rivers running from the ice-front, overloaded with sand and silt, filled up their valleys and ponded back their non-glacial side-streams ; their shore-lines have been briefly described in Ohio and Wisconsin, but the lakes themselves were drained when their flood-plain barriers were terraced ; they form an extinct species, closely allied to the existing Danube and Red River type. As the ice-sheet melts away, it discloses a surface on which the drift has been so irregularly accumulated that the new drainage is everywhere embarrassed, and lakes are for a time very numerous. Moreover, the erosion accomplished by the ice, especially near the centers of glaciation, must be held responsible for many, though by no means for most, of these lakes. Canada is the American type, and Finland the European, of land-surface in this condition. The drainage is seen to be very immature, but the immaturity is not at all of the kind that characterized the first settlement of rivers on these old lands : it is a case, not of rejuvenation, but of regeneration ; the icy baptism of the lands has converted their streams to a new spirit of lacustrine hesitation unknown before. We cannot, however, expect the conversion to last very long : there is already apparent a backsliding to the earlier faith of steady flow, to which undisturbed rivers adhere closely throughout their lives.

Water-surface is, for the needs of man, so unlike land-surface, that it is natural enough to include all water-basins under the single geographic term, 'lakes.' Wherever they occur, — in narrow mountain-valleys or on broad, level plains ; on divides or on deltas ; in solid rock or in alluvium, — they are all given one name. But if we in imagination lengthen our life so that we witness the growth of a river-system as we now watch the growth of plants, we must then as readily perceive and as little confuse the several physiographic kinds of lakes as we now distinguish the cotyledons, the leaves, the galls, and the flowers, of a quickly growing annual that produces all these forms in appropriate order and position in the brief course of a single summer.

W. M. DAVIS.

CAMBRIDGE, MASS., September 7, 1887.

SKETCH OF PYRAMID ISLAND, PYRAMID LAKE, NEVADA.

INDEX.

www.ingramcontent.com/pod-product-compliance
Lightning Source LLC
Chambersburg PA
CBHW021804190326
41518CB00007B/437